AFRICAN
GENERAL SERVICE MEDALS

African General Service Medals

R. B. Magor

The Naval & Military Press Ltd

This revised and enlarged edition of
AFRICAN GENERAL SERVICE MEDALS
first published 1993

Published for the author by
The Naval & Military Press Ltd
Unit 10 Ridgewood Industrial Park,
Uckfield, East Sussex,
TN22 5QE England
Tel: +44 (0) 1825 749494
Fax: +44 (0) 1825 765701
www.naval–military-press.com
www.military–genealogy.com
www.militarymaproom.com

'I cannot conclude without paying a tribute to those brave native troops—Sudanese and Hausas—who help keep up the prestige of the British empire at the farthest African outpost.'

'where in East or West Africa hardly a day passes without a shot being fired in its defence.'

Seymour Vandeleur, D.S.O.

Lieutenant, Scots Guards

19th January, 1898

'Guardian over all stands the Sikh, who being immune to local influence of all kinds constitutes the "motor muscle" of Imperial Authority as he stands erect beside his rifle on guard over British Interests 6,000 miles from the Punjab. He is a picked volunteer from all the Sikh Regiments. If at any time considerations of expense or desire to obtain complete homogeneity in the military forces of the Protectorate should lead to the disbandment of these companies, those who take the decision will have incurred a responsibility which few would care to share with them.'

1906

Winston Churchill

'Whatever happens, we have got the Maxim Gun and they have not.'

Belloc

'On ne fait pas la guerre en Afrique ce qui s'il fait en chasse aux Nègres.'

French Missionaries

Contents

PART II: EAST & CENTRAL AFRICA

Introduction

This book is an extended edition of my notes on the African General Service Medal, first published in 1978.

These notes are about the Small Wars in tropical Africa, for which the undernoted medals with clasps were given:—

Ashantee Medal
Central Africa Medal
East & West Africa Medal
East & Central Africa Medal
Ashanti Star
Ashanti Medal
African General Service Medal

Details are also given on the Emin Pasha Relief Expedition and the Uganda Star.

Over the years I have added detail to the various actions, which I have discovered since the original edition was published.

Much new material is incorporated, most of which has already been published in the Journals of the Orders and Medals Research Society. This includes Henderson's epic action at Dawkita in 1897, a saga comparable with Rorke's Drift in the Zulu War. An account is also given of the Joint Boundary Commission, West Africa, and the actions with the Schütztruppe für Kamerun in 1908—actions for which both the British and German Governments issued dispatches.

I am indebted to Jim Balmer, who has managed to decipher the many illegible notes which I gave him, including further annotations written on the original text.

For convenience this book is divided into two sections, dealing firstly with events in West Africa, and then in East and Central Africa.

As previously, each part gives:—
1. A brief historical background to show how the British came to be there at all.
2. A note on the local colonial forces raised in each territory and the H.M. ships involved.
3. Except in the case of the Ashanti Wars, about which so much has been written, the following résumé:—

(*a*) Extracts from the Army Order or Admiralty Instruction authorising the medal or clasp.

(*b*) A short narrative of where the action occurred, who was involved, what happened and the casualties incurred.

(*c*) Extracts from the Medal Rolls to show very approximately the numbers entitled to the clasp.

Whilst I am very conscious that any work of this nature inevitably remains incomplete, it does serve the purpose of giving some background about the African Medals which are still somewhat sparsely described.

At least the reader can ascertain that Mwele is near Mombasa and not, as I imagined from Gordon's 'British Battles & Medals,' in Nigeria, as the 2nd Northern Nigeria Regt. were said to have been present during this campaign—a fact I have been quite unable to verify.

R. B. Magor
July 1993

Part I

West Africa

CHAPTER ONE

List of Medals and Clasps

Authority, Army Order or London Gazette	Clasp	Reference No. of File	Very approx. numbers issued
	THE GAMBIA		
212/1892	1891–2	W.O. 100/76, ADM 171/45	see note on Date Clasps
7/1895	1893–94	W.O. 100/76	see note on Date Clasps
L.G. 26657/1895	Gambia 1894	ADM.171/46	714
133/1902	Gambia	W.O. 100/76 & 91, ADM.171/356	
	SIERRA LEONE		
212/1892	1887–8 (Yonnies)	W.O. 100/76, ADM. 171/45	349
212/1892	1892 (Tambi & Toniataba)	W.O. 100/76, ADM. 171/45	1250
7/1895	1893–94 (Sofas)	W.O. 100/76	675
	Sierra Leone 1898–99	W.O. 100/92, ADM. 171/48	3000+
277/1906	Kissi 1905	W.O. 100/390	348
	GOLD COAST		
43/1874	Coomassie	W.O. 100/44, ADM. 171/37	–
128/1896	Ashanti Star	W.O. 100/79	–
253/1900	1896–98 (Northern Territories)	W.O. 100/83	–

51/1898	Dawkita 1897	W.O. 100/76	41
249/1901	Kumassi	W.O. 100/93	

SOUTHERN NIGERIA & LAGOS

212/1892	1892 (Jebus)	W.O. 100/76, ADM 171/45	see note on Date Clasps
L.G. 26651/1895	Benin River 1894	W.O. 100/76, ADM. 171/46	456
L.G. 26706/1896	Brass River 1895	W.O. 100/76, ADM. 171/46	417
L.G. 26958/1898	Benin 1897	W.O. 100/76, ADM. 171/46	2500+
253/1900	1897/98	W.O. 100/83	
253/1900	1898 (Hinterland of Lagos)	W.O. 100/83	see note on Date Clasps
253/1900	1899 (Central Territories, Bula)	W.O. 100/83	
234/1902	Aro 1901/1902	ADM. 171/56, W.O. 100/390	1963
133/1902	S. Nigeria	W.O. 100/393	302
129/190 349/1906	S. Nigeria 1902	W.O. 100/393	559
4/1905 49/1906	S. Nigeria 1902–03	W.O. 100/393	570
1/1906 49/1906	S. Nigeria 1903	W.O. 100/393	751
1/1906	S. Nigeria 1903–04	W.O. 100/393	331
1/1906	S. Nigeria 1904	W.O. 100/393	910
128/1906	S. Nigeria 1904–05	W.O. 100/393	386
92/1909	S. Nigeria 1905	W.O. 100/393	78
277/1906	S. Nigeria 1905–06	W.O. 100/393	705
102/1908	West Africa 1906	W.O. 100/394	627
160/1910	West Africa 1908	W.O. 100/394	198
149/1911	West Africa 1909/10	W.O. 100/394	427
460/1924	Nigeria 1918		1000+

Northern Nigeria

155/1897	Niger 1897	W.O. 100/76 and Nigerian Roll	697
253/1900	1898 (Borgu, Lapai, Ibouza, Barua, Anam, Bassema, Siama Illah and Dama)	W.O. 100/83	see note on Date Clasps
253/1900	1899 Bula	W.O. 100/83	see note on Date Clasps
253/1900	1899 (Central Division)	W.O. 100/83	
128/1903	1900 (Munshi & Kaduna)	W.O. 100/92	570
133/1902	N. Nigeria	W.O. 100/391	1350
41/1902 129/1903	N. Nigeria 1902	W.O. 100/391	1039
171/1903 65/1905	N. Nigeria 1903	W.O. 100/392	2174
1/1906	N. Nigeria 1903–04	W.O. 100/392	507
1/1906	N. Nigeria 1904	W.O. 100/392	663
260/1907	N. Nigeria 1906	W.O. 100/392	1372

The East & West Africa Medal

The nine 'dated' Clasps: 1887–8, 1891–2, 1893–4, 1896–98, 1897–98, 1898, 1899 and 1900 are somewhat confusing as the same Clasp was awarded for expeditions in different countries and the date depends entirely on the individual's service time in the area.

Furthermore, the number of personnel on the Roll entitled to the clasps does not indicate the number of Clasps or medals issued, as the Roll shows many present at two or three campaigns for which only one Clasp was awarded; viz. '1892' for Tambi, Toniataba and Jebus, or '1893–94' for Sierra Leone and Gambia, or '1898' for Borgu and Lapai, or '1900' for Munshi and Karduna.

As a corollary, it is impossible in some cases without referring to the Roll to know where the Clasps 1892, 1893–94, 1898 or 1900 were won.

Extracts from the relevant Army Orders and tables showing entitlements are as follows:—

Army Order No. 212 of 1892 stated:—

'The Queen has been graciously pleased to signify Her Pleasure that a medal of the same pattern as that issued for the Ashanti War be granted for operations in Central Africa and on the East & West Coasts.

'Her Majesty has been pleased to approve of a clasp being attached to the said medal on which will be indicated the year or years in which the recipients of the medal were engaged thus.

 (a) For the operations against the Yonnie Tribe 1887–8
 (b) For the expedition up the Gambia 1891–2
 (c) For the expedition against Tambi 1892
 (d) For the expedition against Toniataba 1892
 (d) For the expeditions against the Jebus 1892

'The principle being that the year or years on the clasp cover all the above operations in which the recipient may have been engaged in such year or years.'

Roll W.O. 100/76 & ADM 171/45 show the following entitlements:—

	Yonnie 1887–88	Gambia 1891–92	Tambi 1892	Toniataba 1892	Jebus 1892
The Governor	–	1	–	–	–
Civilians	6	–	2	–	3
Officers, Miscellaneous	8	1	9	2	6
Army Service Corps	1	1	3	1	1
Ordnance Corps	1	–	1	–	–
I.M.S.	3	3	3	2	3
1 Bn. West Indian Regt. inc. details from Depot	298	3	531*	349	61
2 Bn. West Indian Regt.	–	178	–	–	–
West Indian Fortress Co. R.E.	–	–	7	–	–
Gambia Police	–	31	–	9	–
Gold Coast Constabulary	1	–	–	–	150
Sierra Leone Constabulary & Civil Police	44	8	–	–	–
Sierra Leone Frontier Police	–	–	163	–	–
Lagos Hausas	–	–	–	–	169
Naval Personnel ex Acorn, Icarus, Rifleman	38	–	–	–	–
Sierra Leone Colonial Steamer, S.S. Countess of Derby	–	–	23	–	–

 * Approximately 152 personnel of the W.I.R. were present at both campaigns Tambi and Toniataba and 45 at the Jebus campaign.

It is not clear from the Roll, which is headed 'Gambia and Toniataba,' whether the Naval Brigade received the clasp 1891–2 or 1892. As most of those involved only went into action on or after 1st January 1892 the clasp should be '1892' but the evidence indicates that in fact clasp 1891–2 was issued.

Naval Brigade—those who were landed and formed part of the Field Force for operations on shore against Toniataba were:—

H.M.S. Alecto	9
H.M.S. Racer	130
H.M.S. Sparrow	67
H.M.S. Thrush	61
H.M.S. Widgeon	69

These figures included 99 Kroomen. Some Ship's Crews were more original than other over the Kroomen's names, which included 'Bottle of Beer,' 'Peasoup,' 'Teapot,' 'Sodawater,' 'Two glasses,' 'Trouble,' 'Blackman' and 'Prince of Wales.'

Army Order No. 7 of 1895 authorised Clasp 1893–4 ' ... for the following operations
(a) ... against the Sofas ... from 26 Nov. 1893 to 20 Jan. 1894
(b) ... on the Gambia from 22 Feb. to 11 March 1894'
and approved of a clasp inscribed '1893–94.'

W.O. File 100/76 shows the following entitlements:—

	Sofas	Gambia
Officers miscellaneous	–	2
R.E. all Ranks	12	–
Army Service Corps	14	6
Army Medical Corps	3	3
Interpreters	–	2
1 Bn. W.I. Regt. Officers	20	13
Other Ranks	511	372
Sierra Leone Frontiers	138	3
S.L. Colonial Steamer 'Countess of Derby'	22	24

Obviously many men of the W.I.R. served in both Gambia and Sierra Leone.

Army Order No. 253 of 1900 authorised the Grant of the 'West Africa Medal' for operations in 1896–99:—
'H.M. The Queen has been graciously pleased to approve of the West African Medal with a Clasp inscribed '1896–98,' '1897–98,' '1898' or '1899,' according to the date of operations, being granted to the ... Forces who were employed in any of the undermentioned military operations in West Africa ...

(a) In the several operations in the Northern Territories of the Gold Coast between 27 November, 1896 and 14 June, 1898.

(b) In the expeditions in the Hinterland of Lagos between September 1897 and 14 June, 1898.

(c) With the Force under Lieut-Col. J. Willcocks in Borga before 14 June, 1898.

(d) In the Lapai Expedition under Lieut-Col. Pilcher from 8th to 27th June, 1898.

(e) In the expeditions knows as Ibouza, Anam, Barua, Basema and Siama Angiama, Illah, Dama, all in 1898, and Bula 1899.

The medal with Clasp inscribed '1899' will also be given to 'Forces employed in the Central Division Expedition in April and May, 1899.'

Army Order No. 250 of 1901 extended the entitlement under 1(a) and (b) above: ... all Military Personnel employed on duty at any place within the Northern Territories of the Gold Coast or in the Hinterland of Lagos will be considered as entitled to the Clasp. The hinterland of Lagos to include all British Territory between the Northern Frontier, the colony of Lagos proper (about 10 miles from the sea coast) and the River Niger.

File W.O. 100/83 shows the following entitlements:—

	1896–98	1897–98	1898	1899
Commanders in Chief, Governors, Telegraph Bn. R.E., Inf. Regts., etc.	2	50	62	6
Sierra Leone Army Service Corps	–	7	–	–
Army Medical Corps	–	–	10	–
2 West Indian Regt.	209	559	5	–
1 Bn. W.A.F.F.	–	2	348	–
2 Bn. W.A.F.F.	–	23	34	–
Niger Bn. W.A.F.F. (formerly Niger Coast Protectorate Force)	–	–	–	290
1 Bn. N.N. Regt. W.A.F.F.	–	149	68	–
2 Bn. N.N. Regt. W.A.F.F.	–	99	2	–
3 Niger Bn. W.A.F.F.	–	–	–	282*
2 Bn. Gold Coast Regt. W.A.F.F.	–	34	–	–
Gold Coast Constabulary	–	847	–	–
Lagos Hausa Force	2	431	141	2

* The roll shows the entitlements to 3 Niger Bn. W.A.F.F. as 282:—

Central Division   172

Late Claims:—

Ijebu 1892	1	
Benin River 1894	38	
Brass River 1895	14	
Benin City 1897	57	
	110	110
		282

It is assumed that the Clasp 1894 was issued for these late claims.

Army Order No. 128 of 1903 stated:—
'H.M. the King has been graciously pleased to approve of the award of the West African Medal granted by her late Majesty Queen Victoria (A.O. 253 of 1900) being extended to the Imperial and Colonial Forces employed in the Munshi and Kaduna expeditions of 1900 ...
The Medal with Clasp '1900' will be granted to ...
(a) The Munshi Expedition between 4th January and 19th March, 1900.
(b) The Kaduna Expedition between 20th February and 9th May, 1900.'

W.O. File 100/92 shows the following entitlement:—

	Munshi	Kaduna
Officers Miscellaneous	5	10
N.C.O.s	1	1
Indians N.C.O.s	1	1
R.E.	8	–
No. 1 Bty. N.N. Regt.	29	12
No. 1 Bn. N.N. Regt. W.A.F.F.	–	178
No. 2 Bn. N.N. Regt. W.A.F.F.	298	201
R.A.M.C. attached	1	2

Of the 2 N.N.R. there were approximately 142 Ranks on both expeditions and therefore the total number of Clasps or Medals issued for this unit was approximately 350.

THE NAVAL MEDALS

The authority for the various Clasps are in the London Gazette No. 26657 of August 30th 1895.

Witu 1890
Witu August, 1893 (referred to as Pumwani and Jongeni)

       Liwonde 1893 (Chief Liwondi)
       Juba River 1893
       Lake Nyassa 1893
       Gambia 1894
       Benin River 1894

No. 26706 of February 4th 1896.   Brass River 1895

No. 26816 of January 22nd 1897.   Mwele 1895

In 1922 the price of silver rose to such an extent that the silver coinage was worth more as metal than as currency. The Medal drawers of the Admiralty were overhauled and any unclaimed Medals returned to the Mint for melting. Those marked 'Mint 1922' in the rolls were so dealt with.

CHAPTER TWO

The Coastal Areas

'Beware and take heed of the Bight of Benin where few come out though many go in.'

From the very early days the seagoing nations of Europe visited the West Coast and established forts at various points on the seaboard.

The main products traded were slaves, 'oyle of palme,' and gold, in that order of importance. Millions, if not tens of millions, of human beings were transported, traded by their Negro Kings or captors for trade goods, the favourite of which were firearms. 100,000 muskets a year were made for this purpose in Birmingham alone.

Fernando Po, a craggy but very fertile island situated 40 miles from the coast of Nigeria, was discovered by the Portuguese. The papal bulls of 1455 and 1456 granted exclusive rights over the area to the Portuguese and these were reconfirmed by the Pope in 1481.

Queen Elizabeth I granted a charter to Sir John Hawkins to trade with Senegal, Sierra Leone and Gambia. Sir John was knighted in 1562 for supplying slaves to the New World.

Charles II gave a charter to 'The Company of Royal Adventurers' to trade from Gibraltar to the Cape of Good Hope and these were the first real British traders with the West African Coast. The Company of Royal Adventurers was succeeded by The Royal Africa Company in 1672.

The Portuguese established themselves first on the Gold Coast and in 1482 built Elmina Castle, but in 1642 they withdrew from the Gold Coast, having made a treaty with the Dutch over Brazil.

Much later, in 1872, the Dutch also withdrew from the Gold Coast and transferred their settlements at Elmina, Accra, Dixcove and Sekondi to the British.

In 1721 The Treaty of Utrecht formally acknowledged British trading rights in the area.

The first British settlement on the Gold Coast was made in 1650 when they purchased forts from the Danes, built others at Kormantine and then in 1661 built Cape Coast Castle. These were destroyed by the Dutch during 1664–65, but rebuilt by the Royal Africa Company in 1672. This Company was later succeeded by the African Company of Merchants.

Gambia was discovered by the Portuguese in 1447 and in 1588 Queen Elizabeth I granted a charter to certain merchants, who established a trading post there. Our rights were not recognised by the French until 1783 with the Treaty of Versailles.

The Gambia is a narrow strip of land on either side of the lower reaches of the Gambia River, which is navigable by seagoing ships for a distance of 200 miles. The country is mostly mangrove swamps on the edges of the creeks lining the river and thick bush and scrub elsewhere.

Sierra Leone was acquired from native chiefs in 1787 as a refuge for homeless negroes and passed to the crown in 1807.

During the early 19th century Gambia, Sierra Leone and the Gold Coast were united as 'West African Settlements' with their headquarters at Sierra Leone. The Government withdrew in 1828 leaving affairs to private enterprise until 1843 when the Gold Coast and Sierra Leone were again united until 1850, when once more they became separate colonies, jointly administered until 1888.

During the early history of the coast there was much rivalry and struggle for trade, but the British Government did not become seriously involved until the abolition of slavery by International Agreement in 1807, from which time the Royal Navy was stationed on the West Cost of Africa.

The Navy's role was a complicated process of blockading, cutting out operations against slavers' ships, boat expeditions up rivers and the destruction of the slavers' barracoons and warehouses. Several ship to ship actions were fought with the slavers, who as a last resort were prepared to take on His Majesty's ships, as virtually all the slavers' ship were well armed and some had considerably more firepower than the naval vessels with which they fought.

Although England pioneered the anti-slavery movement, it was not until 1811, when slaving was made a felony for U.K. citizens, that the British merchants were deterred from plying this highly lucrative trade which yielded up to three hundred percent profit per voyage.

Other countries also made slave trading 'illegal,' the United States in 1808, Holland in 1814, Spain in 1817 and Brazil in 1829, but their nationals ignored the law and carried on using all the legal loopholes open to them. Indeed, the Governments of the main slave-using countries, the United States, Brazil and Cuba, had really no intention of taking effective steps to hamper the slave trade, as their plantations depended on the supply of slave labour. Whereas the status of slavery was abolished in the British Empire in 1833, the trans-Atlantic trade only ceased after the Brazil slave markets were closed in 1853, and in the United States, ten years later, when Abraham Lincoln rather belatedly proclaimed slavery illegal.

No campaign medals were awarded for these services but they are relevant because the presence of His Majesty's ships necessitated the establishment of bases at places like Freetown, Lagos etc. and the gradual extension of the British presence and 'pax Britannica' into the hinterland by means of treaties with chiefs along the coast, who, in consideration of giving up slave trading and denying their country to foreign slave traders, received a regular salary or subsidy. This in turn led to campaigns and punitive expeditions to suppress slavery, for which the medals and Clasps were given.

In the early days, if they did not receive medals, financially the Navy did well enough from the 'Head Money' they received, as the Abolition Act authorised a bounty of £60 per man, £30 per woman and £10 per child that they set free. This head money was cut to a flat rate of £10 per head in 1814 and £5 in 1830. In addition, the ship capturing a slaver's vessel received a quarter of its value as prize money. In spite of these rewards, however, between 1814 and 1826 only 15,900 slaves were set free by the Navy, to be settled later in Sierra Leone. This was only a tiny proportion of the annual trade, for in one single year Cuba imported 16,000 slaves and Rio de Janeiro 25,769, and by 1816 it was estimated that 60,000 were still shipped annually.

In all, millions were transported across the Atlantic over the years, of which a very high proportion died because of the appallingly crowded and insanitary conditions on board the slavers' vessels.

The West Coast was exceedingly unhealthy and was traditionally called 'the white man's grave.' Malaria, blackwater fever and dysentery were the killers.

How curious the medical principles of Lugard, written in the 1890s, sound today:—

'Avoid all unnecessary exposure to the sun when not in active exercise ... more harm is done by standing in the sun or running out of the tent for a few minutes in the sun with no hat or a small cap than by any number of hours walking in the hottest hours of the day ...

When in hard exercise it does not matter greatly about the quality and purity of the water one drinks but when leading a sedentary life, stagnant or unpure water should be boiled. Removing the hat (to adjust it) in the sun is a folly I see daily perpetrated ... it should be done under the shade of a thick tree. It is essential in tropical climates to protect the stomach, liver and spleen ... wear a thick cummerbund of flannel outside the underclothing ... it is from a chill of these organs that most of the fever, dysentery, diarrhœa and cholera is induced.

He went on to advocate camp fires to dispel the malarious night vapours.

No wonder so many died of sickness.

In 1843 the first British Consulate was established at Fernando Po as Consul to the Bights of Biafra and Benin. The political development of the Nigerias is as shown in the undernoted diagram.

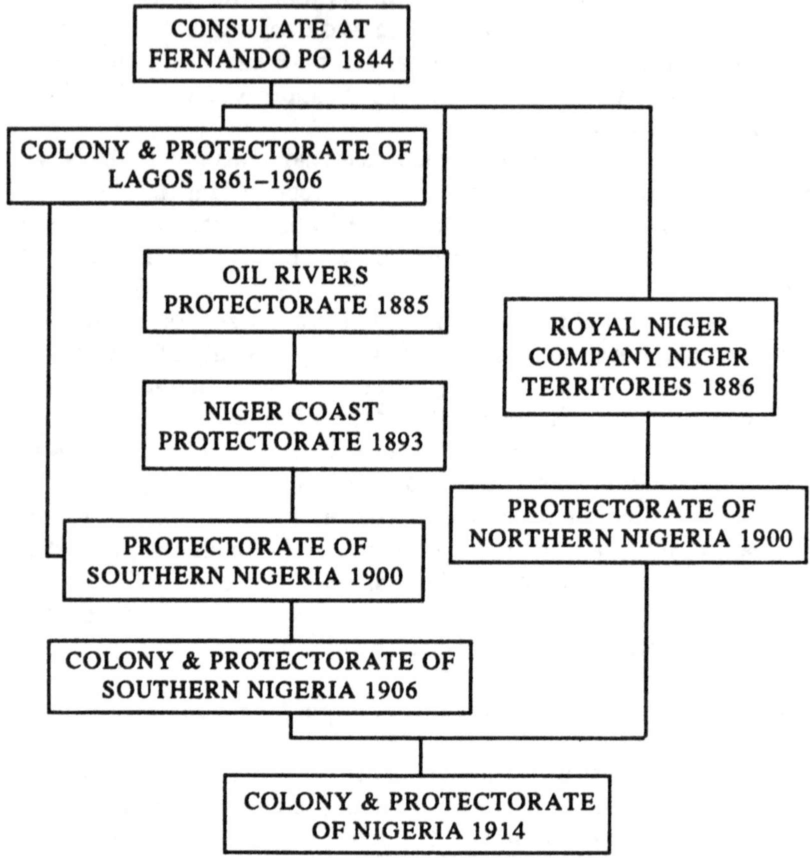

The boundaries of Nigeria are derived from the River Niger and its tributary, the Benue, rather than from any tribal or racial boundaries.

In the 13th century the invading Berbers pushed back the aboriginal negroes to the marshy coastline of the Bights of Biafra and Benin and by the end of the 19th century, Lagos and the two Nigerias had a combined area of some 335,000 square miles and a population o 18,000,000.

The Gold Coast covered 90,000 square miles with a population of 4,000,000.

Gambia and Sierra Leone had areas of 4,000 and 30,000 square miles respectively with populations of 200,000 and 2,000,000.

Unlike East Africa, there is no high mountainous tableland. The Kwahu Plateau between Accra and Kumasi is only 2,000 feet high whilst Lake Chad on the northern boundary is at an altitude of 800 feet only, although the Futa Jallon massif does rise to 5,000 feet. A general description of the area is that it comprises a belt of swampy mangrove bush and thick rainforest extending from the coast for up to 100 miles inland, where it is succeeded by a region of grassland and orchard bush for several hundred miles until it merges into semi-desert on the borders of the Sahara in the extreme northern region.

The Volta River was navigable for only 80 miles, the Niger and Benue for 400 miles or so, whilst at certain times of the year, small vessels could sail up the Cross and Gambia Rivers for 150 and 200 miles respectively.

NIGER RIVER BASIN

Early explorers of the Niger were Mungo Park, who was killed in 1805 at Bussa on the River Niger, and Captain Clapperton and his servant, Richard Lander. The former died in 1827 but Lander returned to the coast in 1830.

The Royal Navy made various forays up the Niger in 1861. Comdr. H. R. Wratislaw with the screw gun-vessel, Ranger (5) and the paddle steamer Brune, conducted an expedition against Porto Novo which was bombarded. Further river trips occurred in 1864, 1865 and 1866 to subdue the more unruly of the chiefs. Vessels used were the Investigator (2) paddle, which drew 4' 4" of water, the Archer (13) screw and Dart (5) screw.

In 1869 H.M.S. Lynx (4) twin screw (Comdr. East) and Pioneer (2) paddle (Lt. W. Wiseman) explored as far as possible up the River Niger. They travelled for 400 miles until both vessels and indeed their crew became 'very sickly.'

No medals were awarded for these activities.

The local inhabitants were very much more diverse than those on the eastern seaboard and had absorbed or been conquered by various other races during the preceding 1,500 or so years.

The National African Company was incorporated in 1882. It received a charter and became the Royal Niger Company in 1899: its charter was revoked on the 9th August 1899 and the territories taken over by H.M.G. for a sum of £865,000. Prior to takeover, the shareholders had generally received a dividend of 6%.

Sir George Goldie, K.C.M.G., was the youngest son of Lt. Col. John T. Goldie-Taubman (Scots Guards). From his first visit to West Africa in 1877 he devoted himself to establishing a British control over the Niger Basin. Although the companies operating in the Niger Basin were making good profits, they had no cohesive policy, and Goldie succeeded in 1879 in welding the four companies formerly directed from Liverpool, Manchester, Glasgow and London into one.

He built up a fleet of gunboats to protect his riverine trade.

The Company received its Royal Charter in 1886 and henceforth had its own judiciary and armed constabulary.

On behalf of the Royal Niger Corporation, Goldie made treaties with the Hausa States and established some forty trading posts on the Niger and Benue Rivers.

He employed Lugard to negotiate treaties with the Bornu kings and thus to deny the north-eastern area to the French who were expanding northwards from Dahomey.

There were early expeditions for which no medals or Clasps were awarded. In 1877, a punitive expedition was sent against Brass and in 1879 Onitsha was shelled by one of H.M. ships. In 1865 the Niger was blocked at Lokoja and in 1871 a British steamer was sunk in the Niger.

CHAPTER THREE

The Local Forces

Each territory established its own Constabulary or Militia which eventually developed into the 'Royal West African Frontier Force.'

During the formative period, one or other of the battalions of the West India Regiment was stationed on the Coast; this unit had British officers and Negro troops from the West Indies. The Navy, too, invariably took part in most of the early expeditions either afloat or as a naval brigade.

Lugard had a low opinion of the W.I.R. because they carried so much impedimenta. It apparently took 10 porters to service one soldier, reminiscent of the dirge from the Kaiser's War in East Africa: 'We are the porters, who carry the food for the porters that carry the food.'

GAMBIA

Sir Charles MacCarthy was Governor in 1815–16, responsible for suppressing the slave trade, and on the 23rd April, 1816 established a garrison of Royal African Colonial Corps on the island of Banjol, which became the site of the present capital Bathhurst.

When in 1888 Gambia was separated from Sierra Leone, its garrison was supplied by the West Indian Regiment.

In 1901, the Gambia Company W.A.F.F. was raised.

SIERRA LEONE

In 1829, the Sierra Leone Police Corps was formed with an establishment of 17 officers, 23 N.C.O.s and 300 African Creoles, Mandes and Timinis. The unit was the forerunner of the Sierra Leone Frontier Police established by ordinance on the 15th January, 1890. This force was known as 'The Frontiers' and was incorporated into the W.A.F.F. as the Sierra Leone Battalion in 1901.

GOLD COAST

Sir Charles MacCarthy organised The Gold Coast's forces with 3 companies of Royal African Colonial Corps of Light Infantry in the early 1820s and at the same time a Militia was formed to prevent raids by the Ashantis. In 1824, the R.A.C.C. was annihilated by the Ashantis at Isamankow where the Governor also was killed, his skull, mounted as a cup, becoming part of the royal regalia.

From 1828–43 only a Militia existed, but a company of W.I.R. was stationed at the Gold Coast.

The Gold Coast Constabulary was raised in 1879 with an establishment of 16 European officers and 1,203 Africans. It comprised the 1st Bn. and 2nd (Northern Territories) Bn., which were amalgamated when the Gold Cost forces were incorporated into the W.A.F.F. in 1901 as the Gold Coast Regiment.

LAGOS

This territory was also garrisoned by the W.I.R.

In 1863 Lt. John Glover (later Sir John and Governor of Lagos) had been exploring the River Niger and was wrecked at Jebba. On returning overland he freed a quantity of slaves which he enlisted and these were originally known as 'Glover's Hausas.' In 1865 they were authorised as a regular force with a strength of 600 and called the Hausa Constabulary which was later renamed The Lagos Constabulary.

This unit was incorporated into the W.A.F.F. as the Lagos Battalion in 1901.

SOUTHERN NIGERIA

When the Oil River's Protectorate was declared in 1885, a force was raised called 'Oil River's Irregulars,' which was nicknamed 'The Forty Thieves.'

In 1891 it was renamed The Niger Coast Constabulary with its headquarters at Calabar.

In 1898–9 the Constabulary became the nucleus of the 3rd Battalion of the W.A.F.F., which became the Southern Nigerian Regiment in 1900.

NORTHERN NIGERIA

McGregor Lair formed the Society of Merchants in the first half of the 19th century and up to 1886 a warship was detailed each year to show the flag.

In 1886 the Royal Niger Company's Constabulary was formed. In 1897–8 it formed the nucleus of 1 and 2 Special Service Corps Battalions W.A.F.F. and was subsequently renamed 1 and 2 Battalions Northern Nigerian Regiment.

THE WEST AFRICAN FRONTIER FORCE

In 1879 the W.A.F.F. was formed, originally comprising 1st and 2nd Special Service Battalions and 3 Batteries W.A.F.F. In 1898 a third Bn. was added. In 1900 The Royal Niger Constabulary was split between the 3 Battalions of the W.A.F.F.

In 1901 the W.A.F.F. was integrated by colonies as follows:—

Formerly

Northern Nigerian Regiment,
   H.Q. Zungeru
     1 & 2 Bn.
     2 Btys. 75 mm. guns
	1st & 2nd Special Service
	 Bns. and Royal Niger
	 Constabulary

Southern Nigerian Regiment,
   H.Q. Calabar
     1 Bn.
     2 Btys. 75 mm. guns
	3rd Special Service Bn. and
	 Royal Niger Constabulary

Gold Coast Regiment,
   H.Q. Accra (later Kumasi)
     1 Bn.
     2 Btys. 75 mm. guns
	1st Bn. Gold Coast
	 Constabulary, 2nd (Northern
	 Territories) Bn. and Gold
	 Coast Constabulary.

Sierra Leone Bn., H.Q. Freetown 1 Bn.	Sierra Leone Frontier Police
Lagos Bn., H.Q. Lagos 1 Bn.	Lagos Constabulary

Gambia Company formed in 1902.

The title 'Royal' was granted in 1928.

The forces of Ghana opted out in 1959 and the R.W.A.F.F. was ended on 1.8.74 when the various countries took over the maintenance of their units exclusively.

Brigadier F. A. S. Clarke in 'The History of the W.A.F.F.' writes:—
'It must also be remembered that, particularly in their early days, units took part in many minor expeditions, sometimes involving heavy fighting and always considerable hardship, for which medals but no battle honours were given.'
C. E. Callwell's textbook on 'Small Wars' gives the African Tribesman small praise:—
'The savage denizen of the tropical forest is, it must be confessed, rather a poor creature as a fighting man and this is a fact which necessarily influences the conduct of operations in bush and jungle to a remarkable extent ... their usual plan is to blaze off a sudden ill-aimed volley and then scatter away to a place of safety.' This description gives no credit at all to the fantastic courage and élan of so many of the tribes who died in almost suicidal attacks on British squares and maxims.

LOGISTICS

A column of 520 Askaris required 125 carriers for a six day expedition, plus 150 for the medical department, 6 men for each hammock. Every maxim required 8 men including ammunition carriers and each 7-pdr. gun 15 men for the carriage and about the same number for ammunition.

If a 75 mm. gun accompanied a column, a further number of 35 carriers for the gun and carriage and about 40 carriers for the ammunition. The 75 mm. guns had to be transported on the heads of porters. The main parts had to be slung on poles fitted with crosspieces so as to distribute the weight equally on the heads of 4 men. The piece, cradle wheels and trail each required 4 men, the axle, 1 man.

THE WEST INDIAN REGIMENT

BATTLE HONOURS: Dominica, Martinique 1809, Guadaloup 1810, Ashanti 1873–74, West Africa 1887, 1892-3-4, Sierra Leone 1895, Palestine 1918, East Africa 1916–18, Cameroons 1915–16.

The Regiment was founded in 1795 by the amalgamation of the Carolina Corps of American Loyalists (1779) with their negro slaves and Malcolm's Rangers of Martinique. The 2nd Battalion was formed from the St. Vincent Rangers and by 1779 there were 12 Battalions but there was a general amalgamation in 1888.

In general the personnel were West Indian negros and all were originally based in the West Indies. In 1824, a depot was opened in Sierra Leone which was to have been used as a recruiting base. Thereafter the Regiments moved between the West Indies and Africa, based at Sierra Leone, Gambia and the Gold Coast.

Recruiting for the 1st Battalion was from Jamaica and the 2nd Battalion from Trinidad and the Eastern Caribbean.

THE WEST AFRICAN REGIMENT

BATTLE HONOURS: Sierra Leone 1898–99, Ashanti 1900, Cameroons 14–16 and Dirala.

Raised 1896 12 coys, HQ 60 BOs, 25 BNCOs, 1500 African Rank & file from Pagan and Muslim tribes in Sierra Leone Hinterland.

Disbanded in 1928 by Army Orders 77 & 78.

CHAPTER FOUR

The Gambia

Army Order No. 212 of 1892 authorised the Clasp '1891–2' for the expedition up the Gambia from the 29th December, 1891 to the 5th February, 1892.

This expedition was against Chief Fodeh Cabbah who had given trouble in early 1891 by resisting the passage of the Anglo-French Boundary Commission through his country when he had attacked the party and wounded several Europeans.

H.M.S. Alecto (4) special service paddle (Lt. F. G. M'Kinstry) was in the Gambia River, and she was joined by the gunboat H.M.S. Widgeon (Lt. G. L. B. Bennett) at Kansala. The Widgeon was a 'Redbreast' class composite gunboat launched in 1889 with a displacement of 805 tons, length 165 feet, speed 13 knots, armed with six 4″ Q.F. guns and two 3-pdrs., and had a complement of 76 officers and ratings.

A small landing party was put ashore with two 7-pdrs. accompanied by His Excellency the Governor himself, together with a party of Royal Engineers.

After an advance to Sangajore, a Chief came and apologised and the expedition withdrew and returned to their ships.

The men had not been able to get out of their clothes once during the 17 days ashore, which in Victorian rig and that climate, must have induced prickly heat to say the least.

However, Fodeh Cabbah continued to give trouble. Commander L. F. Royle of the H.M. sloop Racer, as S.N.O. at the coast, planned an expedition under command of Lt. I. M. F. Fraser (H.M. gunboat Sparrow) to capture Fodeh Cabbah. Fraser and the Superintendent of Police, Captain T. H. Hawtayne, (N. Staff. Regt.) borrowed the Colonial steam launch, Lily, for a reconnaissance, pretended to go off on a shooting expedition, and located Fodeh Cabbah at Marige.

It was decided to make a surprise night attack and reinforcements arrived in the Gambia in H.M. gunboat Thrush and other men of war on 1st January, 1892.

The naval brigade consisted of:—

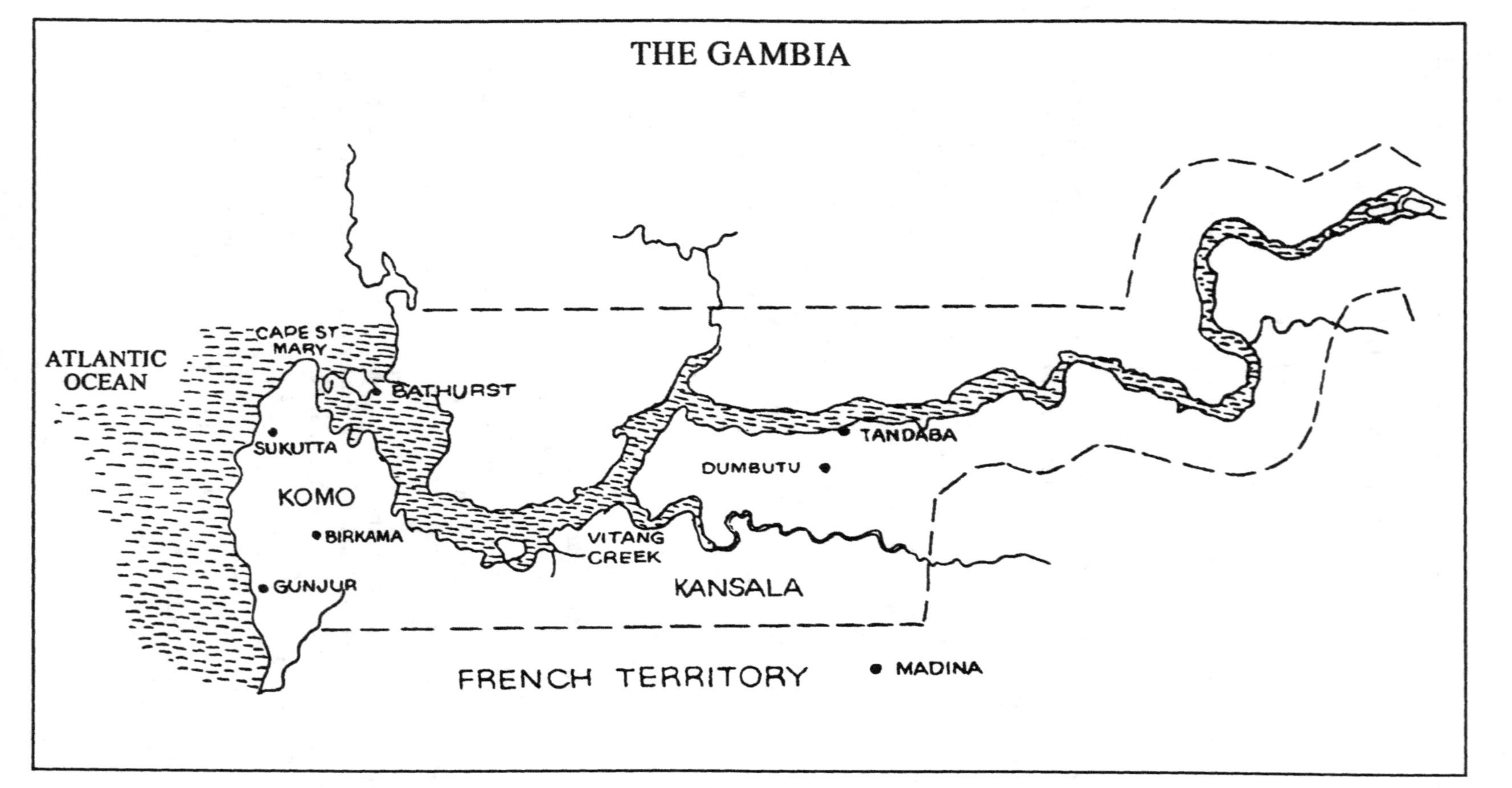

THE GAMBIA
ATLANTIC OCEAN
CAPE ST MARY
SUKUTTA
BATHURST
KOMO
BIRKAMA
GUNJUR
VITANG CREEK
KANSALA
DUMBUTU
TANDABA
MADINA
FRENCH TERRITORY

H.M.S. Racer (Comdr. H. L. F. Royle) 2 officers, 53 seamen, 11 marines and 39 kroomen.

H.M.S. Thrush (Lt. A. J. Loane) 4 officers, 20 seamen, 7 marines and 19 kroomen.

H.M.S. Widgeon (Lt. G. S. P. Gwynn) 2 officers, 20 seamen, 6 marines and 19 kroomen.

H.M.S. Sparrow (Lt. I. M. Fraser) 2 officers, 28 seamen, 7 marines and 19 kroomen.

Total: 13 officers, 140 seamen, 33 marines, 96 kroomen.

Major T. Claridge of the 2 W.I.R. embarked his troops on the gunboats Thrush and Widgeon. After a feint to make Foteh Cabbah think the expedition was to go to Kansala, Marige was surrounded on the night of 2nd January 1892. However, there were obviously insufficient troops for the job as the men were spaced seven paces apart and at 1 a.m. Foteh Cabbah broke out through the cordon on horseback and fled into French territory.

The Naval Brigade then set fire to the various towns and villages which were said to be his strongholds.

A small detachment of 60 W.I.R. was left under Major Claridge to garrison Kaling where a small entrenched post was established. He was attacked and 25 reinforcements were then dispatched from the Thrush under Lt. Gwynn, but as most of the Naval force were suffering from fever, further reinforcements were sent from the Racer and other ships and by this time a mixed force of 400 had been assembled under Comdr. Royle.

There were minor skirmishes with Foteh Cabbah's cavalry and more villages were destroyed.

The Brigade was unable to cross into French Guinea to tackle the enemy and eventually withdrew. This campaign was not in fact a success.

Files WO/100/76 and ADM/171/45 indicate approximately 516 personnel were entitled to this clasp.

Officers etc.	13
Gambia Police	30
2 W.I.R.	168
Naval Personnel	305
Total	516

It is not at all clear from Roll ADM 171/45 whether the clasp 1891–92 or 1892 was awarded to the Naval personnel involved. The roll is titled 'Gambia Expedition and Toniataba' (in Sierra Leone). It would seem that in accordance with para. 3 of A.O. 212 most of the men should have received '1892' as they arrived on the 1st January, but I have a '1891–2' to a P.O. on H.M.S. Racer, which vessel was certainly not on the river in 1891.

The launch Lily does not appear on the roll.

1893–94
GAMBIA 1894

The Army and Navy were awarded different clasps for the same expedition. Full details are given in the London Gazettes dated 28 Feb 1894 and 4 May 1894.

London Gazette No. 26657 of August 30th 1895 (p. 4918) authorised clasp Gambia 1894 *to those officers and seamen and marines who were actually engaged or formed part of the expedition.* This is essentially a Naval medal.

That authorised for the Army by Army Order No. 7 of 1895 was *with clasp '1893–4' for operations on the Gambia from 22nd February to 11th March 1894.*

These clasps were awarded for an expedition against Chief Fodeh Silah, one of Samory's Lieutenants, who for some time had been slave raiding and terrorising the district west of Cape St. Mary and generally threatening the Colony of Gambia.

A naval flotilla was assembled under Rear-Admiral F. G. D. Bedford C.B., consisting of H.M. cruisers Raleigh (Capt. E. H. Gamble), Satellite (Comdr. A. C. Allen), gunboats Magpie (Comdr. H. G. King Hall), Widgeon (Lt. H. Grant-Dalton) and paddle vessel Alecto (Lt. E. L. Lang). Capt. Campbell with a maxim and two rocket trough detachments of the Sierra Leone Frontier Police accompanied the expedition and also some 50 members of the 1 W.I.R. In all, a force of over 700 all ranks.

The plan was for two columns to land, one at Medina Creek and the other at British Combo, and both to advance and unite at Birkama.

The smaller column under Lt. Col. A. D. Corbet (R.M.) of H.M.S. Raleigh consisted of 50 marines and 50 men of the 1 W.I.R. and a field gun. On the 22nd February they marched on Sukutta where they destroyed the stockades suffering only minor losses.

The larger column under command of Capt. Gamble, who was accompanied by the Rear Admiral and the Administrator, consisting of 200 officers and men and a field gun from the ships, marched in the direction of Birkama. It destroyed two stockades but found it impossible to reach Birkama and they therefore marched back to the landing place at Medina Creek. They were followed up by the enemy and sniped at continuously.

While they waited at the creek for the tide to enable boats to approach the shore, Fodeh Silah and his men attacked from the thick cover beside the shore and caused heavy casualties. 4 officers and 10 men were killed, and 6 officers, including the Column Commander, and 40 men were wounded. One quarter of the naval brigade was thus put out of action.

Eventually the boats arrived and they re-embarked with their wounded but left their dead and a 7-pdr. field gun ashore.

Reinforcements of 300 troops from the 1 W.I.R. under command of Major Fairtlough (R.A.) arrived on 1st March with 30 rockets, 2 7-pdr. guns and a maxim.

In the meantime, Lt. Col. Corbet had destroyed Busamballa and then entrenched himself. He beat off an enemy attack for two hours and was then reinforced.

Comdr. C. J. G. Sawle of the Raleigh took over Capt. Gamble's command and with the squadron sailed to Gunjur which was bombarded. A large force of enemy which had assembled to resist disembarkation was dispersed and the brigade landed on the 7th March and advanced, whereupon Fodeh Silah fled.

The naval awards included two promotions, two C.B.s and two D.S.O.s.

P.R.O. File ADM 171/46 show approximately 668 R.N. personnel entitled and some 46 krooboys. The table for dated clasps shows the entitlement from File W.O. 100/76.

GAMBIA

Army Order No. 133 of 1902 authorised the medal with Clasp 'Gambia' to all officers and men who took part in the operations under Lieut-Colonel H. E. J. Brake, C.B., D.S.O., Royal Artillery, in January to March 1901; Bathhurst to be considered as the base for the operations.

With the Ashantee War in progress, the local chiefs refused to acknowledge British Authority and told their people that they were now in charge.

In June 1900 two travelling commissioners, Silva and Sitwell, together with their escort of six constables, were murdered south of the River Gambia at a village called Sandandi. The murderers were known to come from Dumbutu.

An expedition under command of Lt.Col. Brake was assembled with the following forces, 4 companies 2 C.A.R., and 4 companies 3 W.I.R., from Sierra Leone—a total of 802 all ranks. Three H.M. ships, Thrush,

Forte and Dwarf were detailed to cooperate and provide a naval brigade of up to 200 men if required.

The Dwarf was a Bramble class gunboat. She retained a 3-masted rig and her engines gave her a speed of 13 knots. She was 165 feet long, 29 feet broad and drew 11 feet of water. Armed with six 4″ Q.F. guns, she carried a complement of 76 officers and ratings.

As with the former expeditions in Gambia, secrecy and speed were essential if anyone was to be caught as it was such an easy matter for a miscreant to slip over the border into French territory.

The flotilla assembled off Bathhurst on the 10th January 1901 and went up river immediately. At dusk 2 companies of W.I.R. were landed at Vitang Creek with the role of stopping any escape to the north bank. The main force was disembarked at dawn at Tendaba. Every man carried 100 rounds of ammunition and 3 days rations.

It was an 8 mile march to Dumbutu and the force was led by a man called Lan Sanyeng, the son of one of those killed with Silva and Sitwell.

The troops arrived at Dumbutu at noon and obtained complete surprise. The rebels opened heavy fire, but the town with its stockade was taken, two rebel headmen and 60 followers were killed and 200 captured.

Casualties amongst the British were 1 killed and 4 wounded.

This highly successful operation knocked the stuffing out of the rebels and there was little or no more armed resistance, though the troops were involved in much marching from village to village.

It was then decided to carry out a joint expedition with the French against Fodeh Cabbah who was then living in French territory at a town called Madina, which was surrounded by mud walls and garrisoned by slaves as fighting men.

The British Forces were disposed to prevent retreat whilst the French with artillery and cavalry attacked.

On the 23rd March after a three hour bombardment by the French with 80 mm. guns the walls of Madina were breached and the town stormed. Fodeh Cabbah and 150 of his men were killed, for the loss of 2 killed and 7 wounded on the French side.

After the campaign, three of the captured murderers were hanged, and a further five ringleaders deported.

Entitlements:

Civilians etc.	13
Gambia Police	14
C.A.R. (2 K.A.R.)	294
R.N.	33
Bronze Medals (C.A.R.)	2
	356

SIERRA LEONE

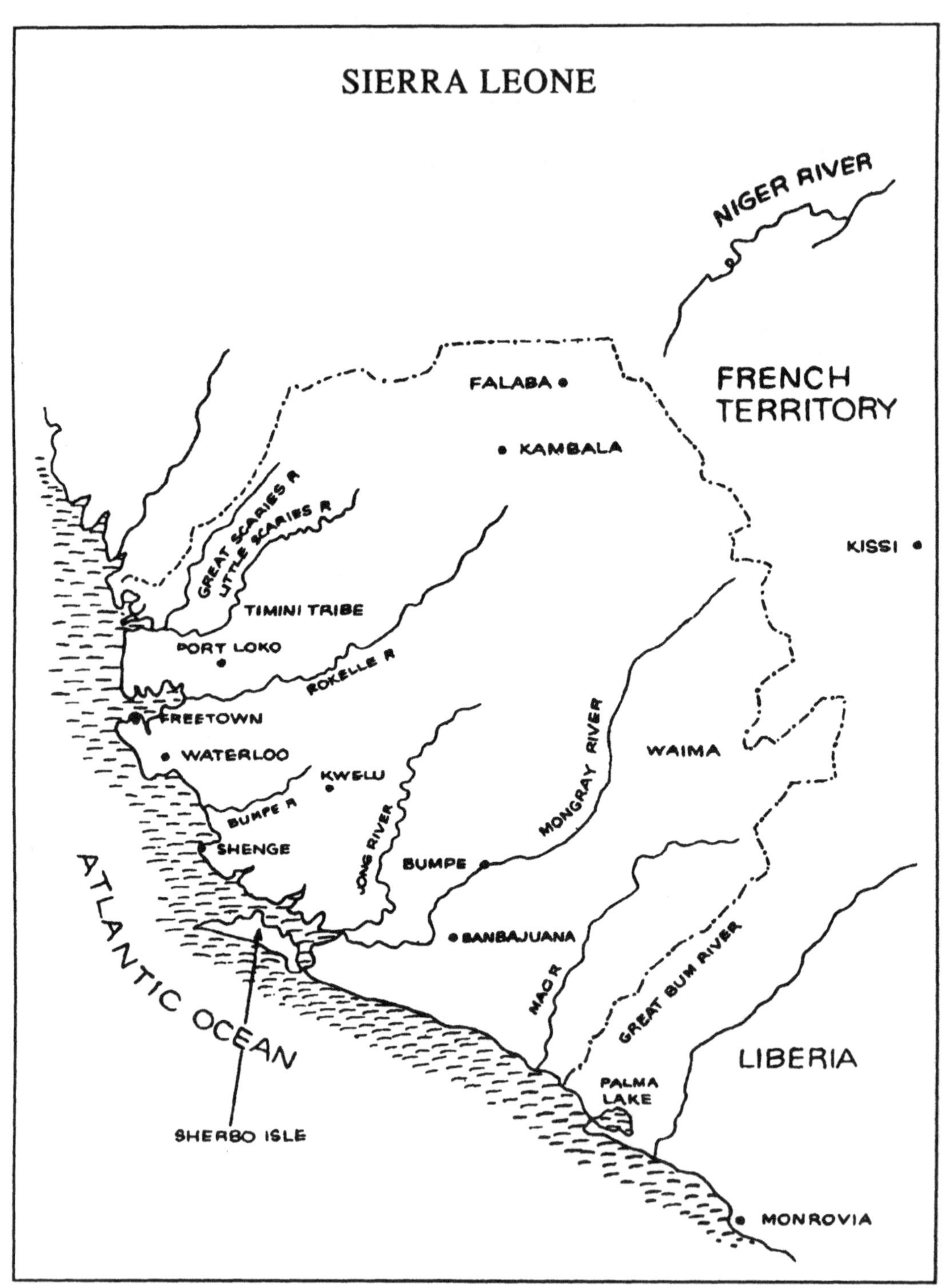

CHAPTER FIVE

Campaigns in Sierra Leone

FOR OPERATIONS AGAINST THE YONNIE TRIBE

1887–88

Army Order No. 212 of 1892 authorised a medal of the same pattern as that issued for the Ashanti War for operations in Central Africa, and on the East and West Coasts for Imperial and Colonial Forces employed in the operations against the Yonnie Tribe from 13th November 1887 to 2nd January 1888, both dates inclusive ... and approved a clasp being attached to the said medal on which will be indicated the year or years in which the recipients of the medal were engaged.

This punitive expedition under command of Colonel Sir Francis Walker de Winton was to punish the Yonnies, a tribe living in the hinterland of Sierra Leone.

The Force consisted of 300 men from the 1 W.I.R., 44 ranks of the Sierra Leone Frontier Police (The Frontiers) and a naval contingent of 15 men from H.M.S. Acorn (Comdr. W. E. B. Atkinson), under command of Lt. F. A. Valentine, who received the D.S.O. for his services. Medals were also granted to personnel from H.M.S. Icarus and Rifleman.

The service was described as extremely arduous owing to the density of the forest which had to be traversed and to the continual fusillade which was kept up by the concealed enemy from their muzzle loaders which fired rough bits of iron and small shot.

During this campaign the Yonnies showed much skill in devising ambush stockades and in making use of obstacles so as to bring the troops to a standstill under their fire.

Robari, the Yonnie stronghold, was eventually reached, shelled, set on fire by rockets and quickly taken. The rebellious chiefs then submitted and the expedition returned to the coast.

British casualties were 20 wounded.

Files W.O. 100/76 and ADM 171/45 show that approximately 400 were entitled to the clasp or medal, but there were some who were already in possession of the 1874 Ashantee medal and so only received a clasp.

Officers, civilians etc.	19
1 W.I.R.	298
All ranks Sierra Leone Frontier & Civil Police	45
Naval Personnel	38
	400

1892

Army Order No. 212 of 1892 authorised the clasp '1892' for the expeditions against: Tambi, from the 8th March 1892 to the 11th April 1892, both dates inclusive, and Toniataba, from the 12th March to the 30th April 1892, both dates inclusive.

In 1891, a Chief called Carimoo (Karimu) had established a stockaded stronghold at Tambi on the River Scaries. He was a notorious freebooter and led gangs of War Boys belonging to Chief Bari Bureh against the tribesmen near the Limba Susu border, who were under British Protection.

In May 1891, Carimoo fired at a party of Sierra Leone Frontier Police and when in March 1892 Capt. Robinson, R.E., and a small force of Frontier Police attacked him, Robinson was killed and his 'Frontiers' repulsed. This defeat needed to be revenged.

Officers were sent out from England and the troops and naval forces assembled.

The military forces under command of Colonel Ellis, C.B. consisted of 550 men of the W.I.R., 125 Frontier Police under Major Moore and 400 friendly Africans, whilst a naval brigade from H.M. ships Racer, Alecto, Sparrow and Widgeon, mentioned in the Gambia 1891–92 expedition, provided a landing party of some 175 seamen. The force was equipped with 7-pdrs., rockets and maxims.

Tambi was assaulted on the 7th April 1892, and after breaches had been made in the walls by artillery, was captured and destroyed.

British casualties were 2 killed and 6 wounded, whilst over 200 enemy were killed as they endeavoured to escape from the town.

The expeditionary force was then reorganised and attacked another stronghold called Toniataba on the 28th April 1892.

Toniataba was well fortified and the enemy under Chief Suliman Santa resisted strongly but eventually the town was captured and destroyed and Suliman Santa was killed.

Lt. H. D. Wilkin of H.M.S. Racer, who led the naval storming party, was awarded the D.S.O.

British casualties included Capt. Roberts, W.I.R., killed and 5 wounded.

From this campaign arose the regimental custom in the Sierra Leone Battalion of singing the National Anthem and reciting the Lords Prayer after Tattoo Roll Call.

Files W.O. 100/76 and ADM 171/45 show the entitlement, and the reader is referred to the dated Clasp table.

Lance Corporal C. W. J. Gordon 1 W.I.R. received the V.C. on 13th March, 1892. His photo appears in Wallis's book, 'The advance of our West African Empire.' He was a West Indian (Negro) although eligibility for the V.C. did not extend to Indian Troops until 1911.

1893–94

Army Order No. 7 of 1895 authorised the Clasp '1893–4' to the Imperial and Colonial Forces for the operations against the Sofas from 26th November, 1893 to 20th January, 1894.

Full details of this expedition were given in London Gazette No. 26490 dated 27th February, 1894, which published the despatch of Colonel A. B. Ellis (1 W.I.R.), Colonel Commanding Troops West Coast of Africa.

The difficulties of the 540 mile march, much of the time on half rations, carrying their wounded in hammocks, and the horrors and devastation caused by the slave traders are vividly portrayed therein, as is the courage and fortitude of the troops involved.

The casus belli was that the Sofas had been raiding from French territory and had devastated considerable areas under British influence. What this meant is aptly described in the despatch, from which I quote:—

'... reached the ruins of Tekwiama. This had been a populous native town of between 3,000–4,000 inhabitants and had been completely destroyed by the Sofas ... In one place were counted 40 headless corpses, with hands and feet tied, thus showing that these men had been made prisoners and then slaughtered ...

'... reached Kerra-Yemma ... here the spectacle met the eyes which exceeded anything yet seen ... on the left hand side before entering the gate in the stockade was a pile of corpses from 7 to 8 feet high. In an examination necessarily brief on account of the horrible effluvia I counted on the surface of the heap the bodies of 27 women, children and infants, and the heap must have contained 100 corpses.

'... to the devastation wrought by the Sofas. This was far greater than I supposed. The Sofas had destroyed all towns and villages north-east of Kommendi for a distance of 15 miles and all those to the west and north-west for a distance of 32 miles ... the operations of the Sofas covered an irregular quadrilateral measuring 75 miles from east to west and about 55 miles from north to south. This region was a complete desert without a single inhabitant. The entire population had been killed or sold into slavery and it will take many years to repair the mischief done.'

At the end of November, 1893, the Adjutant General approved an expedition against the Sofas who were aptly described in the despatch, 'The Sofas under Porro Kerni were simply slave raiders.'

Colonel Ellis sailed from Freetown with his troops and carriers in the colonial steamer The Countess of Derby and the hired steamer Badibou and disembarked at Bende from where they marched 139 miles to Banguma.

The strength here was:—

R.E.	2 officers, 10 NCOs and men
1 W.I.R.	11 officers, 369 NCOs and men
A.S.C.	1 officer
A.M.S.	2 officers and 5 NCOs
Frontier Police	1 officer, 47 NCOs and men
Total	17 officers, 431 NCOs and men with 2 R.M.L. 7-pdr. guns

There were many porters.

Colonel Ellis' intention was to attack Porro Kerni's stronghold of Kerra Yemma, but before marching there, he sent a flying column 270 strong to Yallu (Yelladu). They made a circuitous march during which there were minor skirmishes. The enemy's fire was chiefly from flint-lock muskets and was delivered at too long a range to be effective, for although many men were struck by slugs and fragments of iron, only 8 were wounded.

The columns then concentrated and camped at Waima where before dawn on the 23rd December they were attacked by Lt. Maritz of the French army with a force of 30 African Tirailleurs and 1,200 native allies who thought the British were Sofas. A heavy fire with magazine and chassepot rifles was directed into the camp, but the W.I.R. 'exposed to a storm of bullets from magazine rifles showed wonderful steadiness,' and at dawn the British advanced and cleared the bush to discover Lt. Maritz who was dying. Total casualties were 2 British officers and 8 ORs killed with 16 wounded, whilst the French lost 10 Tirailleurs and many of their allies.

Reinforcements and supplies arrived later that day and after this Colonel Ellis pushed on to Kerra-Yemma (or Kayima). However, shortage of rations forced him to place his troops on half rations as there were no local supplies to be found in the devastated countryside.

The path was exceedingly difficult, leading over boulders or rocks up and down steep declivities, through tangles of roots and creeping plants, and hands had to be used constantly.

On one day's march the troops waded 96 times through swampy streams and 25 times through large stretches of swamp with black, slippery mud up to their ankles or knees. The men's boots and clothing soon rotted and fell to pieces.

In spite of these conditions only one European and two Africans died of sickness.

The deeper, unfordable rivers in their path had either to be crossed by small canoes or on rafts made on the spot from corkwood trees.

En route, Colonel Ellis made a treaty with Chief Sigbanda who said these were the first white men ever to visit his district. It is difficult not to wonder if this Chieftain understood the meaning of 'make any treaty or cede any territory to any other European nation without the consent of Her Majesty's Government for the time being,' when he affixed his mark to the document.

At this time the Sofas attacked the Frontier's small post at Tungea which had just been reinforced and now had a garrison of 48 police. The 1,500 'War Boys' under Chief Vonjo, instead of easily capturing the post, were heavily defeated, and Porro-Kerri, who had accompanied Fori, was killed by a chance rifle shot.

Sub Inspector Taylor, in command of the Tungea post, received the D.C.M.

The Sofas had moved out of Kerra-Yemma but were eventually located after a night's march, attacked and surprised in their stockade at Bagbwema.

They opened heavy fire on the W.I.R. The 7-pdr. guns then shelled the gate, setting fire to the thatch of the houses. Bagbwema was taken with the bayonet, 200 Sofas being killed and 70 captured, whilst a further 70 casualties were inflicted during the pursuit. The British suffered two wounded. 673 slaves were liberated and a large number of muskets and chassepot rifles taken and destroyed. Fori escaped and returned to make more trouble for the French until he was captured years later in 1907.

The expeditionary force then returned to Freetown.

The clasp '1893–94' was also awarded for an expedition on the Gambia and File W.O. 100/76 indicates that the following were entitled to the clasp:—

R.E., A.S.C., A.M.S. etc.	33
1 Bn. W.I.R.	470
Sierra Leone Frontier Police	157
S.S. Countess of Derby	25
	685

The crew of the S. S. Countess of Derby consisted of captain, chief engineer, 2nd engineer, 2nd mate, 7 stokers, 3 stewards, 2 cooks, 1 carpenter, 1 Q. M., 1 coxwain of boats, 2 seamen and 4 kroomen. It does not appear as if they did much to earn this medal.

The steamer was manned by 19 blue jackets and marines from H.M.S. Fox and bombarded 13 villages beside the River Bompe. They then landed and burnt the villages.

SIERRA LEONE 1898–99

Army Order No. 152 of 1899 approved the West Africa Medal with Clasp inscribed 'Sierra Leone 1898–99' being granted to Imperial and colonial Forces who were employed in the military operations in the colony and protectorate of Sierra Leone between 18th February and 9th March, 1899.

London Gazette dated 12th October, 1900 was the authority for the award to naval personnel.

The uprising was in the coastal areas and small forces of R.N. personnel patrolled and fought up the rivers from the coast.

This rising was known as the 'Hut Tax War' and to quote a contemporary writer ... 'was the immediate outcome of a widespread plot, hatched with all the diabolical cunning allied to secrecy which forms so conspicuous a trait in the character of the indigenous African.'

In 1896, when a protectorate was declared it was decided to impose a hut tax of 10/– for a 4-roomed large house and 5/– for a small house, with a view to providing for the cost of the Administration of the country. A contemporary writer states, 'the imposition of the tax caused great discontent among the ignorant natives.' Indeed it did, for in spite of the efforts of the District Commissioners to explain the necessity for the tax, when it became due for payment on the 1st January, 1898, several Chiefs were arrested for refusing to pay. The rest of the Chiefs then declared war and Chief Bai Bureh of Kasse was the leader of the centre of resistance. There is no doubt that arrangements were carefully made for a coordinated rising with the objectives of getting back their country, reintroducing slavery, abolishing the Hut Tax, driving out British rule and the white man and murdering everyone who could speak English or wore European clothes.

There were three ringleaders, the paramount Chief Niagu of Paguma, paramount Chief Guburu of Bompe, who was later hanged, and Bai Sherbo of Yonni.

Major Tarbett of the Sierra Leone Frontier Police reinforced Captain Sharpe, the District Commissioner, but although they patrolled in Bai Bureh's country and destroyed his stockades, they could not catch him. Tarbett then marched to Karene where he was practically surrounded by the rebels and 3 companies W.I.R. were required to keep open communications with Port Loko.

The rebellion spread quickly and large numbers of English speaking Bandajuma were murdered with savage brutality. Many were captured before they had an opportunity of even trying to escape. The case of Mr. Hughes of Imferri is horrifying even to those who are accustomed to the atrocities of the modern world. He was kept tied up and daily brought to trial and 'fined.' The fine took the form of cutting off his fingers, then his hands, all this in the presence of his wife, eventually after days he died and then Mrs Hughes' throat was cut. A daily removal of part of the body was a favourite method of murder.

The Governor, Sir Frederick Cardew, asked the Navy to convey reinforcements. Ships available were 2nd class cruiser H.M.S. Fox (Capt. F. H. Henderson) and H.M.S. Alecto (Lt. A. F. Holmes) and 3rd class cruiser H.M.S. Blonde (Comdr. Peyton Hoskyns). This force conveyed a coy of W.I.R. and relieved Major Norris who was is Lagaar and sorely pressed. Indeed, if it had not been for the Navy's timely arrival terrible atrocities might have resulted and the British Colony might have been overrun by the rebels.

Lt. Col. Marshall took command at Karene, cleared the road to Port Loko and caused the withdrawal of Bai Bureh, who was finally captured in October by the W.A.R. under command of Colonel E. R. P. Woodgate.

In April, the Chiefs of Rokella openly declared war, but the District Commissioner, Captain Fairtlough, with a force of 50 Frontiers patrolled the area and kept the insurgents in check. His base at Kwellu was. attacked and Col. Woodgate took a column and defeated the Mendes who were attacking. They regained control of the district by rigourous patrolling.

To the south-east Captain Eames had 50 Frontiers at Bandajuma and Captain C. B. Wallis 28 at Kambia on the Small Bum River. The latter place was attacked and Wallis had to withdraw to Bonthe as his ammunition was exhausted. Colonel Cunningham arrived there in May with reinforcements and proceeded up the Jong River to Mafwe, a trading and missionary post which had been sacked and most of the Europeans killed.

Here a force of 1,000 Mendes attacked him and they were driven off with 130 casualties. In a subsequent attack a further 40 were killed.

Cunningham then advanced on Bumpe where two stockades were captured. These stockades were made of logs up to six feet high backed by boulders and rocks to a thickness of four feet which were impervious to 7-pdr. shell or rifle bullets. It is absolutely impossible for European eyes to discern them by any outward sign; occasionally an exceptionally quick-sighted native would discover the location of a stockade by some indication such as a dead twig or leaves drooping overhead. Generally, the first thing anyone knew was a volley fired at a few yards range which inevitably caused casualties particularly amongst the officers of the W.I.R.

In May the naval forces were reinforced by 1st class cruiser H.M.S. Blake (Capt. Alwinsloe) and 3rd class cruiser H.M.S. Phœbe (Capt. R. S. Rolleston) and H.M.S. Tartar (Comdr. J. T. White).

The rains then set in and the campaign was brought to a conclusion by marches throughout the territory by six separate columns of troops who finally crushed resistance.

The tactics used by the Army were for the column to march in single file through the bush and when opposition was met, the sections faced outwards and fired volleys blind into the undergrowth. The expenditure of ammunition, known as 'burning powder,' was extremely high so the supply of ammunition often ran short. In one reconnaissance to Okemug the column had to retire and eventually reached camp with only 3 rounds left out of the 100 carried per man. Nowadays presumably a chopper gunship would be called up to give neutralising fire.

Two auxiliary units apparently took part in this campaign, the Sierra Leone Volunteers and the Waterloo Volunteers. The former was commanded by Major J. Walsh (Lt. R.N.R.) with an establishment of two captains (both from the Manchester Regiment), the Mayor of Freetown as an Honorary Captain, two Lieutenants who in civilian life were the Assistant Director of Public Works and the agent for Elder Demster & Co. and an Honorary Surgeon.

For NCOs there were six sergeants and two corporals; their civilian occupations included the Curator of the Botanical Station, a volunteer from the West Indies and the Native Government Officer.

There were 111 privates, some 19 apparently were Europeans including the General Superintendent of the Weslyan Missions and a Frenchman who had left the colony when the roll was compiled. The rest were natives including Sir Samuel Lewis, C.M.G., member of the Legislative Council, and later Chief Justice of Sierra Leone.

The Waterloo Volunteer's roll is less informative, and the only remark was that one member of a Lt., 1st Glamorganshire Volunteer Artillery, This unit's name was apparently derived from the Waterloo District where it operated, which is a few miles south-east of Freetown.

In all, some 4,000 officers and men were employed in the expedition. Imperial Troops lost 4 officers and 17 men killed and 16 officers and 94 men wounded. The Sierra Leone Frontier Police lost 46 killed and 4 officers and 72 Africans wounded.

The despatch stated: The conduct of the troops under fire was excellent. Although they knew the enemy was sheltered behind stockades, which were proof against shellfire and nearly so against rifle fire, they nevertheless fearlessly faced the enemy's fire with a steadiness which would have been creditable on parade.

Files W.O. 100/92 and ADM 171/45 show the entitlement. The latter really is a virtual duplicate of the W.O. File.

The London Gazette dated 12th October, 1900 authorises the clasp to the personnel of H.M. ships Alecto, Blonde and Fox who during the year 1898 were actually landed for military operations or who took part in the boat expeditions under fire.

These include:—
1   River Lokko, 5th March,
2   Sherbo District, 1st–15th May,
3   Boom Kittan River Expedition, 16th May,
4   Up Bompe River, 11th–14th May.

File W.O. 100/92

Civilian Awards: District Commissioners, Surgeons, Schoolmaster and miscellaneous civilians	35
Constables	10
Government Boatmen	19
Interpreters and Court Messengers	45
Miscellaneous Officers	9
Personnel Royal Artillery including Africans	177
Royal Engineers	56
Army Service Corps including labourers	45
Army Medical Service	44
1st Bn. West Indian Regiment	1,123
2nd Bn. West Indian Regiment	594
3rd Bn. West Indian Regiment	159
West African Regiment	895
Sierra Leone Volunteers	126
Waterloo Volunteer Corps	61
The Sierra Leone Frontier Police Corps	553
Colonial Steamer: The Countess of Derby	19
Naval contingent, including 37 kroomen	
H.M.S. Blonde (Less 1 deserter)	124
H.M.S. Alecto	34
H.M.S. Fox (87 Medals, 8 Clasps)	95

KISSI 1905

Army Order No. 227 of 1906 authorised the medal with Clasp 'Kissi 1905' to all officers and men under command of Captain (local

Major) C. E. Palmer, Royal Artillery, who took part in the operations in the Kissi country between 27th March and 28th June.

Chief Kafura in the south and Chief Fassalokoh in the north had been raiding from French territory in Liberia into British territory to capture slaves and Kafura had even attacked the Sierra Leone Battalion's outpost at Korumba.

Liberia's consent to an expedition to punish the slavers was obtained and Major Palmer's expedition, consisting of 15 officers and 359 other ranks of the Sierra Leone Battalion W.A.F.F. and 600 carriers, concentrated at Korumba. Kafura's town of Kenema was their objective and the force first occupied the stockaded camp at Boandu and the towns of Fankissi and Sefadu. Using the latter place as a base, two companies went out on patrol, were attacked, and fought a two-hour battle with the Kissi at Komendi. The tribesmen withdrew when they were counter-attacked by Captain Le Mesurier (Royal Dublin Fusiliers) and Lt. Murray (East Surrey Regiment) of the rearguard. Komendi was then burnt.

This punitive expedition then followed the usual course of burning towns and villages with the occasional skirmish. However, Chief Kafura had had enough and fled into exile into French territory.

After a refit at Korumba, Major Palmer turned his attention to Fassalokoh in the north and marched through very heavy rains to Kaduma which he used as a base for operations. Fassalokoh and his neighbour Dimba fled, but later they attacked the column's camp near the French frontier at We. The attacks continued at intervals by night and day and Lt. Haseldine was wounded.

The expedition the marched southwards and at Wulade a fortified post was constructed and a palaver held to appoint a new Chief in place of Kafura. This ended the expedition which was just as well as over half the officers were sick and unfit and the troops too were very shorthanded.

Capt. Palmer was awarded a D.S.O.

W.O. 100/390

Member of the British Consulate Monrovia:—
    Charles Braithwaite Wallis, H.M.'s Consul to Liberia
    Corpl. John Coulson, Messenger at British Consulate
    Harburuall, Inspector at Consulate
    Tucker Moses, Police Constable   4
British Officers & B.N.C.O.s   18
    Lt. L. P. Reeves, S.L., deceased medal issued to mother
    Dr. J. A. Scotland, Colonial Medical Dept.

Major C. Palmer
Capt. F. N. LeMesurier
Capt. M. M. Bond
Capt. L. Murray
Capt. C. C. Norman
Capt. A. Farrar
Lt. J. G. V. Hart
Lt. J. S. Burra
Lt. H. E. Bailey
Lt. E. von Brockdorff
Lt. W. H. Supple
Lt. H. Goodwin
Lt. R. K. Masaldine
Lt. C. W. C. Crooke, deceased
Dr. W. F. Campbell
Dispenser George Boyle

Sierra Leone Bn. W.A.A.F.	326
Total	348

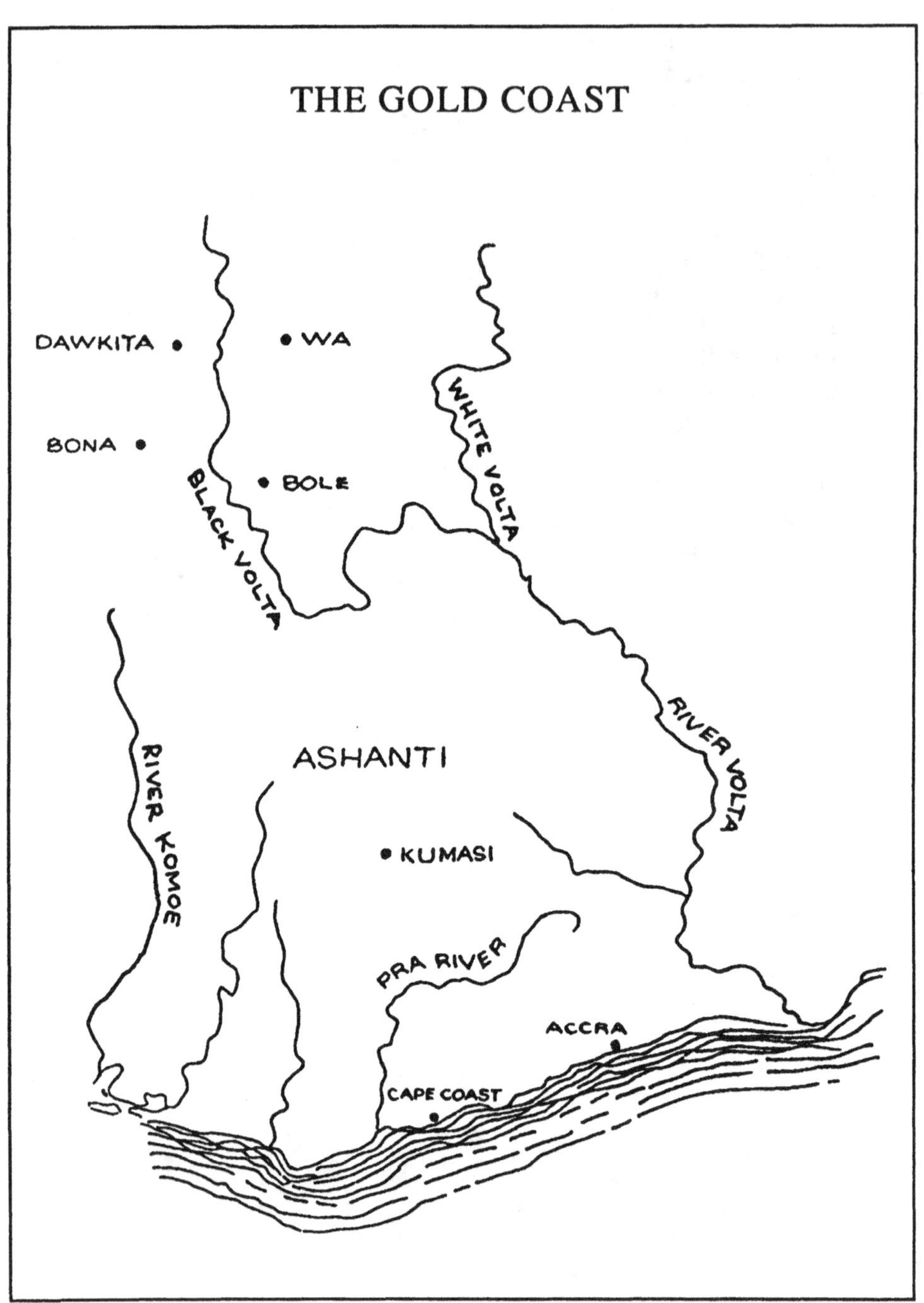
THE GOLD COAST
DAWKITA
WA
BONA
BOLE
WHITE VOLTA
BLACK VOLTA
RIVER VOLTA
ASHANTI
RIVER KOMOE
KUMASI
PRA RIVER
ACCRA
CAPE COAST

CHAPTER SIX

Gold Coast

Ashanti Wars

The three medals granted for campaigns against the Ashantee Tribe are the subjects of numerous books of which Mr. Alan Lloyd's 'The Drums of Kumasi,' published by Longmans in 1964, gives a very excellent account and is readily available from Public Libraries. Details of those entitled to these decorations are shown in Gordon's 'British Battles and Medals.'
The Authorities for the three medals are:—

1. General Order No. 43 of 1874 authorised:—
'a silver medal be granted to all Her Majesty's Forces who have been employed on the Gold Coast during the operations against the King of Ashantee,-with a Clasp in the case of those who were present at Amoaful and the actions between that place and Coomassie (including the capture of the capital), and those who, during the five days of those actions were engaged on the north of the Prah in maintaining and protecting the communications of the main Army ... officers and sailors who during that period or for any portion of that period were on board ship on the coast.

2. Ashanti Expedition 1895–6—Grant of Star
The authority for the Ashanti Star is Army Order No. 128 of July 1896, which approved:—
'the grant of a special star, cast in gun metal, to the Imperial and Colonial Forces employed under the command of Major-General Sir F. C. Scott, K.C.B., K.C.M.G. in the operations connected with the Ashanti Expedition between the 7th December, 1895 and the 17th January, 1896.

3. Ashanti Expedition 1900—Grant of Medal
Army Order No. 249 of 1901 stated:—
'His Majesty the King has been graciously pleased to approve of a new medal being struck to commemorate the operations in Ashanti, which were undertaken in consequence of the rebellion of the native tribes and the siege of Kumassi.
The medal will be granted to:—

55

... all Imperial and Colonial Forces who were at any time attached in orders to the Ashanti Field Force between 31st March, 1900 and the 25th December, 1900, both dates inclusive.

... The British Gaman detachment of the Gold Coast Constabulary under the command of Capt. G. W. S. Oden 3 Bn. Royal Munster Fusileers, during the period of the above operations.

A clasp inscribed 'Kumassi' will be granted

(a) to all who took part in the defence of Kumassi between the 31st March, 1900 and the 23rd June, 1900 or remained in the fort until 15th July, 1900.

(b) to all who started with the relieving column from Bekwai on the 13th July and reached Kumassi on the 15th July, 1900.

(c) to all who started with the column under the command of Colonel C. A. P. Burroughs from Bekwai on the 4th August which reached Kumassi on the 5th August, 1900.

1896–98

Army Order No. 253 of 1900 authorised the Clasp '1896–98' for several expeditions in the Northern Territories of the Gold Coast between the 27th November, 1896 and the 14th June, 1898, and this was extended to all military personnel by Army Order No. 250 of 1901.

Following the Second Ashantee War (1895–96) it was necessary to consolidate the British position in the Northern Territories, particularly because of the French and German influence in this area.

Vandeleur in his memoirs writes that when he was at Accra, 'Small columns were then being sent out beyond Ashanti in all directions to try and occupy some of the Hinterland which was rapidly being parcelled out between the French and German expeditions.'

I have been unable to find any other details and it would seem that only 200 or so of the W.I.R. were eligible. This clasp is therefore rightly described by Gordon as 'rare.'

DAWKITA 1897

Army Order No. 51 approved the medal for operations on the West Coast of Africa with a Clasp inscribed 'Dawkita 1897' being granted to Lieutenant F. B. Henderson, late Royal Navy and to the force employed under him in the defence of the town of Dawkita in the Gold Coast Hinterland.

The Defence of Dawkita by Lt. Francis Berkeley Henderson, R.N., Rtd., who quite wrongly described himself as 'unfortunately I am no soldier,' is one of the more heroic episodes against overwhelming odds in British military history in Africa, yet it is dismissed in most books with a paragraph—Gordon's 'when it was attacked by Sofas,' or Hayward & Clark in The History of the Royal West Africa Frontier Force 'The Sofas attacked him with 8,000 men.'

Lt. F. H. Henderson was born in 1859, the fourth son of a parson, married to the daughter of a Rear-Admiral. He entered the Navy in 1874, was promoted Lieutenant in 1882 and invalided in 1887. In 1895 he was appointed P.S. and A.D.C. to Sir William Maxwell, Governor of Gold Coast Colony; he accompanied him to Kumasi on the 1895/6 Expedition and received the Ashanti Star. In 1896 he was appointed a Travelling Commissioner of the Gold Coast Colony and was later Chief Commissioner of the Ashanti, being awarded the C.M.G. in 1902 (London Gazette 26 June 1902) as District Commissioner in Ashanti. He rejoined the Navy in the Kaiser's War, became a Commander and died in 1934.

How does the Dawkita action compare with the defence of the Post at Rorke's Drift, the epic of 1879?

The British at Dawkita were outnumbered by 175 to 1 against 15 to 1 at Rorke's Post, the enemy lost 400 or 300 approximately in each action and the defenders' casualties proportionately were much the same, Rorke's Drift was twelve hours fighting, whereas the siege at Dawkita lasted for 4 days and nights. The fighting at Dawkita, however, was not hand to hand and the Sofa tribesmen did not match up with the Zulu warrior, as Henderson himself wrote:— 'They are great cowards these Sofas and will bolt before a decent sized force, more especially if it is well supplied,with artillery.' Nevertheless the Sofas were a 'regular' native army and better armed than the Zulus, few of whom carried firearms.

Henderson's Force included an African Surveyor called Mr. George E. Ferguson, a Native Police Officer Mama Gimalah from Timbucktoo and 41 Constables from the Gold Coast Constabulary.

My grandfather Major William Boycott, late 29th Regt., who farmed at Hawkestone in the Karkloff Valley of Natal was appointed Colonial Commandant of District No. III (Pietermaritzburg) during the Zulu War.

His duties included the organisation of what nowadays we would call the Home Guard. I have his copy of the Natal Colony Colonial Secretary's exhortation to all the Colonial Commandants after Rorke's Drift requesting them to explain to their Districts how 'practical it is for a few. determined men to repulse the attacks of large bodies of savages if advantage is taken of cover.' An account of the action pointed out that some 80 men of the 2–24th Regt. assisted by a few others held off an impi of 2,000 Zulus throughout one night, killing more than 300 Zulus and suffering 13 killed and 9 wounded themselves and 'that the gallant defence of Rorke's Post shows how a small party of Europeans, well armed, and behind some sort of entrenchment, may with confident hopes of success protect themselves from an apparently overwhelming number of Kaffirs.'

No less than 11 Victoria Crosses were won on 22 January 1879 at the defence of Rorke's Post.

Henderson was besieged for 4 days and nights by a well armed army of Sofa tribesmen under command of the son of Alimani Samory, the Mandingo Chief, estimated at no less than 7,000 men, of whom 400 were killed at a loss of 2 killed and 9 wounded to the British. Although there was talk of recommending Henderson for a Victoria Cross and the Queen herself much hoped that Mr Henderson's gallant conduct would meet some honorary recognition as 'Her Majesty was much impressed with his story and has known him personally for a long time as a brother officer of the Duke of York in the Navy,' he only received the D.S.O. This was gazetted on 8 March 1898 in the War Office list of the London Gazette in recognition of the services in the conduct of certain military operations against the Sofas in West Africa whilst holding the appointment of Travelling Commissioner.

The background to the Dawkita Siege was that in the 1890s there was competition between the British, French and Germans to increase their possessions in Africa. Feelings ran high indeed in 1896, the Fashoda Incident on the Nile had almost precipitated a war between France and England.

The French were established in the Ivory Coast and the Komoe River was roughly the boundary between the spheres of influence of the British and French Colonies. Although the British were established on the Coast at Accra and Cape Coast and recently at Kumasi, their hold on the Hinterland was tenuous as it was dominated by Alimani Samory who carried out savage raids in the areas of both French and British influence. His modus operandi was to lay waste the areas he invaded, kill all the men and enslave the women and children. His followers did not cultivate the land they occupied but obtained their supplies by conquest. There were complaints, too, that the French themselves acted contrary to the Brussels Act and had caused devastation to many villages in the Mossi area, and had traded in slaves and murdered the inhabitants. The Germans were supposed to be supplying Samory with arms and ammunition.

It was the practice of the European Powers to enter into Treaties of Friendship and Protection with the Chiefs or Kings of borders of the areas in their spheres of influence thus extending their territories. Henderson was sent to Bona to conclude a treaty with its King.

The British treaties were standard Colonial Office printed documents and their provisions included:—

'Whereas the King and Chiefs and Principal Headmen have presented to the Governor of Gold Coast Colony a request that their country should be placed under the Protection of Great Britain and have agreed to enter into a Treaty with Great Britain.

The King and chiefs hereby place themselves under the protection of Great Britain declaring that they have not entered into any Treaty with any other Foreign Power.

The Governor of the Gold Coast Colony hereby takes the Country under the protection of Great Britain.

The King agrees they will not enter into any war or commit any act of aggression on any chiefs bordering their country by which trade shall be interrupted or the safety and property of subjects of H.M. The Queen be lost, compromised, or endangered.

Her Majesty's Government will respect the habits and customs of the country but will not permit human sacrifices and slave dealing.'

The treaties, therefore, placed obligations on both sides.

The Colonial Office wished to extend the British sphere of influence to the West to include the Bona area adjacent to the Gold Coast Colony. Henderson, therefore, proceeded to Dawkita where, on 16th March, he entered into a Treaty with Kampa who had been placed on the skin (elected King or Chief of Bona) only two days previously. The Governor's despatch stated that:—

'The new King of Bona, whose predecessor was killed by the Sofas, is not yet in possession of his town which was occupied by Sofas. The Treaty was accordingly made at 'Dokita' in Lobi Territory.'

The Union flag was hoisted and doubtless in view of the murder of his predecessor by the Sofas, King Kampa must have been relieved to be placed under the protection of Great Britain. Henderson hoped to be able to occupy Bona shortly and re-establish the King in his capital and it was agreed that H.M.G. should then tell France that this area was now claimed by Britain.

Henderson's position was precarious: the reinforcements he expected, Captain Cramer and 50 men, had not arrived at Wa, the Sofas were said to be starving and very short of food and his occupation at Dawkita blocked their way to forage for food and it may have compelled them to attack him.

Henderson wrote to the Sofas 'if you continue to molest our people your action will be considered a hostile one,' and he reminded Samory that 'the King of Ashanti, thinking he was powerful, ignored our warnings and he and his principal men are now prisoners in Elmina Castle.' Samory who claimed Wa by force, replied: 'Go back the other side of the Volta at once, if you want to be killed by war stay at Dawkita.'

Nevertheless Henderson was hoping that with reinforcements he could drive the Sofas out of Bona 'if the necessary men can be spared from the colony.'

Two days later Henderson heard the Sofas were on the march and he made preparations for his defence against attack.

His appreciation of the situation was:—

'The Sofas at Bona are very short of food and must soon get some or evacuate Bona. I am between them and their food and to get it they must attack me. I have a warlike and plucky race around who I feel sure will back me up well. If I leave for Lobi in 2 or 3 weeks the Sofas will have eaten or spoilt all the food and destroyed a fine race as they kill all the men and carry off the women and children. We have provisioned three of the native compounds. They make excellent forts. The walls are about 10 feet high and there is but one small entrance. The flat roofs of the rooms inside are about 7 feet high so you have a parapet of nearly 3 feet—a perfect defence. The compounds were about 70 yards apart.

Our coming here has put the Lobis in good heart and they are eager for the fray. Unfortunately I am not a soldier but it seems that we can defend this part of the Colony best by remaining in Lobi. I have ordered Capt. Cramer to come here with his 50 men, leaving one gun at Wa. I consider, then, that this force here, together with the Lobis, will be able to render good account of themselves should the Sofas attack us.'

On the 29th of March it was reported that the Sofas army was nearby and expected at Dawkita the next day at daybreak.

Henderson therefore gave orders for watering and manning the 3 compounds, sent a runner to Capt. Cramer who was overdue at Wa, told the King of Bona to send all his non-combatants out of town to the Volta River, which he failed to do, and sent out messengers to raise the Lobi.

Henderson's Despatch was written on his return to England and sent direct to the Colonial Office on Sir William Maxwell's instructions, for he was so ill when he reached Accra in June that he was unable to write any report then.

It deserves quoting:—

'Each compound was held by 13 Hausas, No. 1 being under my command, No. 2 under that of Mr. Gimalah, native officer, and No. 3 under Mr. Ferguson. Their orders were, as the supply of ammunition was limited (120 rounds a man) that only picked shots were to fire unless a volley was necessary to prevent a rush. The bush had already been cleared to a distance of nearly 150 yards from the compounds.

At 12.30, the Sofa army appeared over the hill in front of the town, massed in a huge square, numbering, as far as we could estimate, about 7,000 men. Their mounted men, of whom they had about 400, tried to pass to the rear of the town on our left, but were driven back by some Lobis, and then, passing along the river bank, occupied the water-hole distant about 350 yards from our front. The Bonas, who were to have held this, made no resistance. I felt very much the want of a 7-pdr. gun, as, with its assistance, I could have kept command of this water-hole, want of water being one of the chief causes of my having, eventually, to retire.

At 4.30 p.m. their riflemen (numbering over 1,000) opened a heavy fire on us from the cover of the bush. They showed some skill in skirmishing and their fire was well directed, though high. I had been informed on good authority that they were very short of ammunition, but this want had evidently been supplied from some source, as they were able to keep up an almost un-intermittent fire through that night, the next day and the best part of the next night, and at intervals heavily for the rest of the four days. At night I only replied to their fire by an occasional volley, when they came to the edge of their cover.

On the 31st March after three hours heavy firing, there was a sudden cessation about 10.30 a.m., and shortly a man was seen approaching, carrying a letter in a cleft stick. It stated that they were Kalfura Sofas—did not wish the English, who were their friends, to die of hunger and thirst, as their quarrel was with the Bonas and Lobis, who had made the mischief by their lies, and that we had better go away over the Volta River. We thought this showed that the Sofas were disheartened by their heavy losses and inability to get food from the town. From the messenger, a captive Bona, we gathered that the Sofas had lost heavily, were disheartened by our resistance, quite unexpected, and were likely to retire. In reply I informed the Sofas that it was not my wish to right them, but they had attacked me and should retire, as I should not, and was expecting reinforcements from several parts of the Hinterland.

After desultory shooting the enemy began heavy firing about 3 p.m. and continued till dark. That evening we held a council of war. We had heard that no help would be given us by the Lobis, we had no news of Captain Cramer, had only one day's water left, and were running short of ammunition. We decided that unless help came we would try to fight our way through to Wa the next night. We chose the night for this movement as we were aware the Sofas were most reluctant to move or fight by night, and subsequent experience showed us that this was the case. We also decided to inform Capt. Cramer, if possible, of our intentions, which we did by means of the Lobi Chief, Honaki. I ordered that each man, if possible, should have 30 rounds with him, and that the old cartridge cases should be destroyed. The specie, £100, was to be sent to me to be made up into £10 packets for trusty men to carry.

The Bonas who had not already escaped now began to give themselves up, and soon the town was empty, and my pledge to assist the inhabitants no longer bound me to occupy the town longer than I deemed expedient. At 4.30 p.m., after heavy firing, compounds 2 and 3 were in flames and their garrisons retired to my compound, losing in the short passage one killed and six wounded, one mortally. Among these was Mr Ferguson, who had a bullet through the lower part of the leg, which had, however, not touched the bone or arteries. These fires were said to have been caused by Bonas. The fires soon went out and the compounds were occupied by the enemy, who opened a heavy fire from which a small wall on my roof, acting as a traverse, fortunately shielded us, only one man being wounded. Our misfortunes seemed to embolden the enemy, who now approached to the edge of the cover and from all sides opened a tremendous fire with rifles and Dane guns.

To this I replied with an occasional volley when a rush seemed probable. At dusk the firing suddenly ceased, and we decided that our chance had come as it was expected the Sofas would now go for their food. We could not leave by the entrance as this was commanded by No. 2 compound, and had to descend from the roof, a height of nearly 14 feet by a crude ladder which was not long enough—it was consequently difficult to get Mr. Ferguson down. Our formation was a hollow square with the wounded and carriers in the centre, Mr. Gimalah taking direct charge of the men, leaving me to look after the wounded and supervise the withdrawal.

The Government tent, used as an awning and riddled with bullets, I had to leave as well as all my own property. The flag, which had been flying since my occupation of the place, I took down and it was carried to Wa.

After we had advanced 50 yards, the Sofas shouting 'the white man is off to Wa' opened fire without effect, and a straggling fire was opened from 100 yards ahead of us, but two good volleys cleared the way. We were making our way under the direction of a Lobi guide to a path to the left of the ordinary road, said to be unknown to the Sofas, to reach which we had to cross very rough ground. The night was very dark and the enemy seemed to be misled at to our route, as there was no real pursuit, though a great noise was made in the Sofa camp and fires lit in all directions. It was necessary repeatedly to halt and reform the column, and once I found Mr. Ferguson missing, my hammockmen having thrown him down and left him in a panic while crossing the bed of a small stream. He came in shortly, walking with the help of a carrier, before I had started to look for him. I then had him carried in a native hammock, walking behind him and threatening to shoot the bearers if they attempted to bolt.

After about four miles of this we reached the path, which we followed in single file, getting food at a village at 3 a.m. and reaching the Volta about 8 a.m. At the village Mr. Gimalah reported that the bugler, rear face of the square, 13 men and a corporal, were missing. As I understood they had a Lobi with them I did not feel very anxious about them and these men all eventually came in.

After a halt at Wenchi, where we heard further firing to the eastward, we reached, after nightfall, a village distant 8 miles from Wa. There we remained for the night being exhausted by our march of over 40 miles. Early next morning we met Captain Cramer with fifty men, two guns and rocket troughs. We decided to retire to Wa which we reached about 11 a.m. My loss out of 43 combatants at Dawkita amounted to 2 killed and 8 wounded. The enemy's loss we estimated at about 400 and several horses.

We proceeded to get in water and provisions, but though there was food in the deserted town it was hard to induce some of the timid carriers to fetch it. I sent for the King of Wa and asked for horsemen to act as scouts, doubting whether the Sofas were following me from Dawkita. The King was out of his wits with fear, saying that all his men had run away as the Sofas were approaching in force. He then left and I heard he had fled down the Daboya road.

On the 4th April we received information that large bodies of Sofas were close to the town. After consultation it was decided to retire from Wa before we were surrounded, as our retirement in that case would involve heavy loss and the carriers numbering nearly 400.

The necessary orders had hardly been given when we heard that a body of Sofas had occupied a water-hole to the West, about FL mile distant. During their absence a considerable number of the enemy was seen passing to the north of our position towards the Daboya road, which we had previously heard was occupied by them, and soon afterwards Sofas occupied all the compounds near us and opened fire. It was evident that we were surrounded, and we therefore decided to wait till night time, as we knew the disinclination of the Sofas to fight then as confirmed by my experience at Dawkita.

About 4 p.m. we heard a great drumming, which the Sofa prisoner— whom I had brought from Dawkita believed to announce the arrival of the Jimini army under Samory's second son Molai, whose forces had drums, while the Bona army had horns. It occurred to me that if I could see the Prince in command of the Bona army, I might be able to persuade him to agree to a cessation of hostilities. I knew the Prince to be both treacherous and cruel, but considered the risk I ran in going to his camp to be worth incurring if this course could prevent the inevitable slaughter of carriers and loss of guns and ammunition.

I then laid this proposition before the other officers, who at first opposed the plan on the grounds of the risk I should incur, Ferguson especially saying that I should be uselessly courting death in some unpleasant form. In the end they concurred and a letter was written to the Prince stating that we had not come here to fight his people and could not understand why he had followed me here after I had left Lobi and that he might like to see me now and arrange matters although he had refused to see me at Buale. The letter was taken to the camp by an Ikoranza carrier at about 5 p.m. The firing had now practically ceased. About 6.30 p.m. the messenger returned accompanied by a Sofa bearing a reply to my letter. It stated that the Prince wished to maintain friendly relations with the English, and did not wish to fight us. The bearer added that orders had been given to cease firing. I wrote again to say that I would visit the Prince in the morning.

The next morning a chief came to the British fort and after discussion it was agreed that Henderson should go to treat with the Sofas. Ferguson again tried to dissuade him but then bravely volunteered to accompany an interpreter on his mission which he considered was suicidal. Before Henderson left it was distinctly understood that if the Sofas detained him the fort was to be evacuated that night.

Henderson found the Prince surrounded by all his chiefs and young courtiers and behind them about 1,000 riflemen.

The palaver commenced and after the usual preliminaries Henderson said he had not come to fight the Sofas but to prevent the French from occupying the country. He confirmed he wished to march to Daboya en route for Kumasi. Just as things seemed to be about to be agreed the Prince's priest created trouble and he was told that all his men and stores had to come into the Sofa camp and he must write the requisite order, which Henderson refused to do. He however wrote to Cramer to tell him the result of the palaver and assumed that if the palaver failed and he was detained, the British Force was to make their way that night to Daboya.

Henderson was then put under guard and his messenger returned with some food and Cramer's suggestion that he should offer £1,000 for his release; this Henderson would not do but offered the sum for a free passage for all the force, which was declined.

The Sofas threatened Henderson, who, alone and powerless, must have had a particularly anxious time. He was told to wait while his fate was decided and after a wait of 1½ hours was brought before the Council, when he was told:—

'that they had examined my head (figuratively of course) and had seen that I was a good man and would therefore do me no harm but send me down to the Coast but I was a big man and must see the Alimany in Jimini first.'

The next day the Sofas moved to Wa where he found the Headman of the Carriers, George Watson, bound.

In the afternoon Henderson was called to see a head just brought in, which was that of Mr. Ferguson.

'I need say nothing of my feelings at this abrupt information of the death of a friend, whose loyalty, courage and intelligence I had so fully learned to appreciate.'

An eye witness said that Ferguson, who had been wounded and was deserted by his hammockmen, was found in the morning by 2 Sofas at whom he pointed his empty revolver. These ran away and returning with others shot him and then cut off his head.

On 18th April the Party left on horseback to visit Somary at Jimini and on the 26th reached the Komoe river and three days later arrived at Hara Monkoro, Samori's seat of Government, where he met Samori and his elders and 3–4,000 riflemen formed up in an oval. The spoil, guns, etc., captured at Wa was carried around in a kind of procession, followed by Henderson in his hammock. He refused to kneel before Samory but shook his hand instead and then sat down on a chair beside him and a review of the troops took place.

He was then sent to comfortable quarters with an orderly to preserve him from intruders.

Several conversations were held with Samory in which he laid great stress on his friendship with the English and did not consider that either were responsible for the conflict at Dawkita, saying 'it came from god.'

After more talks Samory, who was intensely suspicious, decided to send Henderson to the Coast. He left on the 4th of May with a letter for the Governor and he left accompanied by 16 of his people and a Sofa escort. Henderson by then was ill with dysentery and in a somewhat critical condition owing to a poor diet and lack of any medicines.

Eventually Henderson arrived at Kumasi and after 4 to 5 days he had recovered sufficient strength to be carried to the Coast.

His despatch states that:—

'Mr Gimalah, native officer, was of great service by acting as interpreter on many occasions and he could not speak too highly of the courage and ability displayed by him during the defence of Dawkita.'

He also mentioned Cpl. Mam Zampara and L/Cpl. Malarun and also brought to notice:—

'the admirable manner in which the Hausas who were with him at Dawkita conducted themselves—no troops could have behaved better under fire and their discipline was all that could be wished. Your Excellency is aware how any recognition of such service is valued by the whole Force. I therefore venture to suggest that should the rules of the service permit the West African Medal might fairly be awarded them.'

Of George Ferguson he said, 'as a coloured man he was one in a million' and ends his despatch:—

'... the irreparable loss the colony has sustained by the death of Mr. George Ferguson. From the circumstances in which we were placed and my constant intercourse with him made me fully realise his great ability and thorough knowledge of the country and its people. His untiring devotion to duty and readiness to give any assistance in his power, together with his courage in emergencies and high personal qualities I shall never forget. I have lost a valued friend and the colony an invaluable servant.'

The Colonial Office Files show how they pondered on the reasons for the attack and opined the most probable explanation was that the Sofas were so short of food owing to Henderson preventing their raiding that they had practically no alternative but to remove the British.

The desirability of Henderson's action in visiting Samory was also debated. it was regarded as rash but done in the hope of saving the Bonas and Lobis under British protection from being raided and destroyed. He showed great pluck and good judgement in the fighting and his going alone for a palaver with the Sofas was a most courageous act. He ought to get either the C.M.G. or the D.S.O. or possibly both? For Mr. Gimalah the D.C.M. was suggested. It was suggested that the Hausas receive the West African Medal but Sir Redvers Buller of the War Office had advised 'a medal is never granted for an isolated engagement but only for a campaign or expedition.'

The War Office, however, were of the opinion that 'some order of distinction ought to be instituted for military operations under a colonial or protectorate Government.' The military knew how to deal with recommendations for the D.S.O., etc., coming from their own officers but they could not estimate the relative value of recommendations from civilians, e.g. Sir H. Johnston, Governor of British Central Africa, to whom they especially referred. It would no doubt be convenient if the Colonial Office should have something equivalent to the D.S.O. and the Medal for Distinguished Conduct for the grant of which the

Secretary of State for the Colonies could be responsible but this would take time to arrange. It was therefore recommended that the West African Medal be granted for Mr. Henderson's Force, the D.S.O. for him and the D.C.M. for Mr. Gamalah.

It was minuted that Mr. Henderson's pluck in his behaviour to Somary may have saved his life, it was a bold act to shake the old man's hand instead of kneeling down and no doubt told on these savages.

The Colonial Office authorised Pensions to Mr. Ferguson's widow and children.

The Medal Roll for Dawkita was compiled in Accra by an Inspector of the Gold Coast Constabulary on 18 September 1898. It does not include Mr. George Ferguson nor some of the individuals mentioned by Henderson in his correspondence with the Governor prior to the Siege, nor all those individuals mentioned in his official despatch:—

	N.O. Mama Gimalah	Died at Yabouan
1239	L.S. Mama Gajera	Hinterland
1641	L.O. Osumanu Moshi (5)	Hinterland
1361	L.C. Badentia Grunshi	Hinterland
1592	L.O. Allasan Bichi	Died in Action
601	Pte Aboki Grunshi (1)	Lost in Action
1758	Pte Alhandu Grunshi (2)	Hinterland
1511	Pte Awandu Goriko	Hinterland, Wounded in Action
1863	Pte Alandu Bagamuri	Hinterland
1490	Pte Aboki 'Grunshi (37)	Hinterland
1470	Pte Neti Grunshi	Hinterland
1786	Pte Kaniba Kanjarga	Hinterland
1519	Pte Manandu Grushi (2)	Hinterland
1092	Pte Arawago Moshi	Died in Action
1948	Pte Ekra Kanjara	Died in Action
1652	Pte Jimalah Grunski (1)	Died in Action
808	Pte Jidado Fulani	Hinterland, Wounded in Action
375	L/C Dogo Yaro (3)	Hinterland
1936	Pte Bila Moshi (5)	Hinterland
	L/C Nassamo Wakala Wakala	Deserted the Force
1802	Ag Drmr Pong Grunshi	Hinterland
1363	Pte Belasani Moshi	Hinterland
1278	Pte Bela Moshi (2)	Hinterland
1839	Pte Braima Wongara (4)	Hinterland
1851	Pte Alakarimu Salagna	Lost
1848	L/C Malash Gakabi	Hinterland

1486	Pte Mama Bazarimi	Hinterland
97	L/S Kaku	Lost in Action
1313	G.R.L.S. Kofi Grunshi	Lost in Action
1762	Pte Amadu Balashina (2)	Hinterland
1413	Pte Quarrama Grunshi	Hinterland
1795	Pte Musa Bazarimi	Hinterland
1240	Pte Mahama Bakanu (1)	Cape Coast
	L/C Suleman Bakashina (3)	Kumasi
1578	Pte Bukari Fulani (2)	Kumasi
1291	Pte Zakare Zozo	Kumasi
1426	Pte Suleman Fulani	Kumasi
1667	Pte Samba Fulani	Kumasi
552	Pte Dambari	Accra
1309	Pte Osumano Bakano (2)	Accra
—	Dispenser Oko	Accra

Neither Capt. Cramer, nor Capt. Haslewood, who joined Henderson at Wa, nor their troops were entitled to the Dawkita Medal as it was granted solely to those who were present at the Siege of the Fort.

The medal to No. 1592 Allasan Bichi who 'died in action' is named with large impressed capitals.

Sir William Maxwell died on board ship of malarial fever off the West Coast on 14 December 1897.

It will be seen from the Roll that N.O. Mama Gimalah died shortly after the Expedition. Alimani Somary was captured by the French and died in captivity in 1900.

This must be a very rare medal as it does seem very unlikely that the medals of the majority of constables shown as being in the Hinterland have survived. It rarely comes on the market. It was sold recently at £1,600 which makes it cheap compared to a Rorke's Drift at £5,000.

CHAPTER SEVEN

Southern Nigeria

1892

Army Order No. 212 of 1892 authorised the Clasp '1892' to the Imperial and Colonial Forces in the expedition against the Jebus from the 12th to 25 May 1892, both dates inclusive.

The Ijebu are a tribe living around Ijebuode, north-east of Lagos. Their policy of keeping their roads closed to foreigners hindered British expansion and in 1891 a treaty to open their country was forced on them.

The Governor Carter therefore obtained permission for a punitive expedition if the Jebus closed their country again, which they shortly did.

The expedition under command of Colonel C. F. Scott, C.B., comprised 55 men of the W.I.R., 344 Lagos and Gold Coast Hausas and a levee of 100 Ibadan warriors and 343 carriers under 11 officers, with 3 7-pdrs., 2 Nordenfeldts and Maxim. They embarked at Lagos, sailed down the Lagos Lagoon. The flotilla consisted of the Government Yacht, tugs, steam launches and canoes. Although Epe is only 30 miles from Lagos, no European had ever been over the road. The expedition arrived at Epe on 12 May and 186 more carriers were collected.

During the battle Capt. Owen, Lawrie and Hardinge were wounded, also 5 killed and 40 wounded. The Jebus confessed to losing 17 chiefs and 1,000 men killed. The town of Magbon was the occupied without resistance.

There was heavy fighting in the thick forest and a 22 hour battle at the crossing of the Yemoji River. The British rockets, maxim guns and 7-pdrs. were however more than a match for the Ijebu's sniders and Ijebuode was occupied on the 20th May after one week's campaign.

F. D. Lugard in his book 'The Rise of Our African Empire' wrote: 'On the West Coast, in the 'Jebu' war, undertaken by the Government, I have been told that 'several thousands' were mowed down by the maxim.'

The Medal Rolls W.O. 100/76 and ADM 171/45 show the entitlement, and the numbers of troops involved in the Jebu campaign are shown in the table for the dated Clasps.

NIGERIA

SUDAN
(FRENCH SPHERE OF INFLUENCE)
ARANGU
SOKOTO
MORADU
KATSINA
BARUA
SATIRU
GANDO
KOTORWASKI
THABA RIVER
MAIDAGURI
JEGA
HAUSA STATES
KANO
GOMBA
BEBETI
GUJBA
BURMI
ZARIA
TELWA
KADUNA
BAUCHI
GOMBE
KILDA
BUSSA
ZUNGERU
ABUJA
OKUTA
KADUMA RIVER
GONGOLA RIVER
FORT GOLDIE
OBIDA
KEFFI
WASE
YOLA
JEBBA
RIVER NIGER
GERKEWA
ILORIN
EGGA
RIVER BENUE
LOKO
IBI
ANGLO GERMAN AGREEMENT OF 1890
OYO
IFE
KABBA
LOKOJA
ABINSI
ABEOKUTA
AJAOKUTA
OWO
KATSINA
ILARO
UBIAJA
IDDA
LAGOS
IJEDOODE
EPE
EBU
BENINCITY
ABBOR
ENUGU
ABABKALIKI
BENIN R.
ASABA
CROSS RIVER
OGWASHI-OKU
ONITISHA
OBURRA
KWALE
BENDE
UGEP
WARRI
OGUTA
FORCADAS
OWERRI
AROCHUKO
ITU
CALABAR
BIGHT OF
BENIN
AKASSA
BRASS-BONNY
BIGHT OF BIAFRA

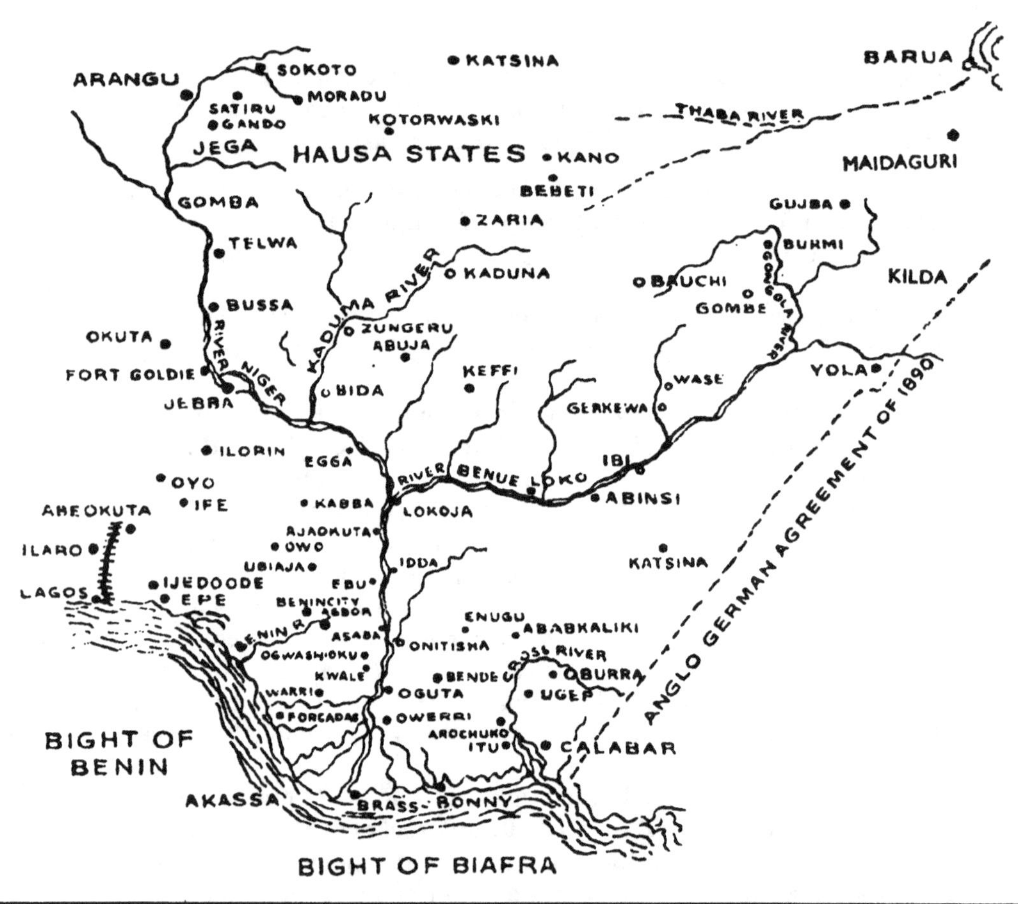

BENIN RIVER 1894

THE EXPEDITION AGAINST CHIEF NANNA OF BROHEMIE.

The Admiralty Despatch dated 20th December, 1894, giving details of this expedition, was published in London Gazette No. 26581 of the 21st December, 1894.

The authority for the award of the medal and Clasp is given in London Gazette No. 26657 of August 28th, 1895 and is quoted in full under 'Witu 1890.' The medal was issued to those officers, seamen and marines who were actually engaged and formed part of the expedition on the Benin River (Chief Nanna, August and September, 1894).

Chief Nanna, who lived in Brohemie town on Brohemie Creek, in a low-lying fœtid area of mangrove swamps on the Benin River, was a powerful chief. For a long time he had terrorised the surrounding countryside for upwards of 100 miles, and by force of arms had compelled almost all the local trade to pass through his hands.

His forces were well-armed as his armoury was extensive including 106 cannons of 2″ to 6″ calibre with 14 tons of gunpowder, 100 swivel blunderbus guns, 1,546 flintlocks, case shot, a gatling gun, etc., etc. Nanna, who was Governor of Benin, had 5,000 slaves in his stronghold and used his position to carry out slave raids.

The Administration enlisted the help of Lt. Heugh of H.M.S. Alecto, who met with a reverse and the Consul-General for the Niger Coast Protectorate had to telegraph for further naval assistance. H.M.S. Phœbe (Captain F. Powell) arrived at the Benin River where she joined H.M.S. Alecto.

A large number of natives collected at night opposite his ship, and Lt. Heugh burnt the three villages at the entrance of the creek. He then took Alecto's steam-cutter, which had been completely armour-plated against rifle fire, on patrol with two officers from the Niger Coast Protectorate and met an obstruction in the creek. On turning they were fired on at point blank range by a gun from a masked battery on the bank, and were holed in the stern between wind and water by shot weighing between seven and nine pounds. They suffered casualties; the two Protectorate officers were disabled, the coxswain killed and two others wounded. The Chief Gunner's Mate, R. H. Crouch, fired a rocket into the battery which silenced it. Steered by Heugh and driven by a wounded stoker, J. Perkins, the launch limped back in a sinking condition with her Nordenfeldt gun disabled and out of action and with the armour plate wrecked. Crouch and Perkins were awarded the C.G.M. and Heugh the D.S.O.

Lugard, who was on his way to Borgu, had been invited to join this expedition but declined as he had fever. It was, in his opinion, in any case a somewhat light-hearted affair, and how right he proved to be.

The Phœbe and Alecto then bombarded Nanna's stronghold and sent a gang of Africans to cut a road to the town passable for artillery.

On the 27th August, a force was landed consisting of 144 officers and men from the Phœbe, 35 from the Alecto, with 157 men of the Niger Coast Constabulary and 80 native woodcutters. They had two RML 7-pdrs., two maxims and the Alecto's rocket apparatus and advanced in open square formation. A few shells were fired into the stockade which was then rushed, but the defenders had fled. Twenty-three guns of 3″ and 4″ calibre were captured, dismounted and spiked, and the advance continued.

The going was very bad with soft, stinking mud. A deep creek had to be filled with branches before it could be traversed and the force then came under fire and a sailor was wounded. As it was late, it was decided to return to the ships, but the tide had risen and the Phœbe's 7-pdr. had to be spiked and abandoned in the creek. It was dark by the time the force re-embarked.

Rear Admiral Bedford then arrived with the cruiser Philomel (Captain C. Campbell) and gunboat Widgeon (Lt. H. Grant Dalton).

Nanna's town had been shelled on the 19th September and on the 20th two parties were landed. Captain Powell, accompanied by the Admiral, commanded a brigade consisting of 136 officers and men of the Phœbe, 35 from the Alecto, 50 marines with a maxim, rocket tubes and a gun cotton party and 100 Constabulary. Captain Campbell's brigade consisted of the Philomel's men in their boats.

Brohemie was occupied without opposition and Nanna fled. He was almost caught when he tried to escape by night in a canoe, leaving behind £327 in cash, his cap and coat and two hideous carved wooden arrangements described as 'the War Juju of Nanna and Bizanne.' The expedition captured Nanna's treasure, stores and war canoes. They destroyed 106 guns besides a great deal of trade goods, which included 'Gin, cases containing 12 bottles in each, 8,300.' How it must have upset the Navy to destroy this last item! The Phœbe's 7-pdr. was also recovered.

The difficulties of campaigning on the West Coast were described in a military manual, 'It is sickness and not the loss involved by actual conflict which saps the strength of a regular army ... Hostilities often take place in unhealthy and even deadly climates in torrid, fever-stricken theatres such as Dahomey and Benin, when this is so, troops are decimated by ill health even when the war is of short duration.'

The Admiral received the K.C.B., and Captains Campbell and Powell the C.B.; in addition, three D.S.O.s and two C.G.M.s were awarded.

File No. ADM 171/46 shows that clasps were earned by H.M. ships Alecto, Philomel, Phœbe and Widgeon. Approximately 451 R.N. personnel and 50 kroomen were thus entitled.

Other sources indicate that 150 Coast Constabulary were also recipients of the clasp. File W.O. 100/76 shows five officers also entitled.

BRASS RIVER 1895

In the London Gazette No. 26706 dated February 4th, 1896, the Department of the Accountant-General of the Navy Admiralty gives notice that *'the medals and Clasps for the Brass River Expedition February 1895 are now ready for issue to those officers, seamen and marines who were actually engaged in and formed part of the expedition.'* The medal was described and the method of application procedure was similar to that for 'Witu 1890.'

The traders at Brass, followers of King Koko, were jealous of the activities of the River Niger Company's trading post at Akassa and on the 26th January, 1895 paddled over in a number of canoes and attacked and destroyed the factory, massacring or capturing all who were there.Forty-three African captives were mutilated, butchered and eaten. A missionary recognised one warrior who had been educated in England painted white and dancing around naked adorned with bits of human flesh. The Acting Agent General and a dozen or so white men however managed to escape.

Rear-Admiral Sir Fredrick Bedford, in command of the Cape and West Africa Station, led the punitive expedition against Koko. The following ships were involved: H.M.S. St. George, flag (Capt. W. C. C. Forsyth), Barrosa (Comdr. J. L. Mark), gunboats Widgeon (Lt. H. Grant Dalton) and Thrush (Lt. H. L. Tottenham). The landing party consisted of 150 Blue Jackets and marines and 150 Niger Coast Constabulary and the flotilla proceeded up the Brass River on the 20th February, 1895.

The gunboats Thrush and Widgeon opened hostilities by shelling the bush on the shore near Nimbi. The pinnaces and launches, each with a machine gun on board, sailed into Nimbi Creek and landed the brigade at Sacrifice Point.

Koko's men attacked in their war canoes which were engaged by the launches' machine guns. When three of the canoes had been sunk, the rest retreated.

On the 21st February stockades near the shore were blown up. The following day the two gunboats sailed further up the creek and begun to shell Nimbi, while the boats advanced on Koko's town. As they neared the town a masked battery opened fire on the boats, scored several hits and caused casualties. However, the boats pressed on, landed, stormed the stockades and captured Nimbi which was burnt.

Casualties included Lt. G. T. Taylor of the H.M.S. St. George, who was in the leading boat, and two seamen killed, with 5 Blue Jackets, 2 Constabulary and 2 kroomen wounded.

On the 24th February, the Barrosa and Widgeon bombarded Fish Town which was destroyed by a landing party the next day.

Most of the Chiefs who had been concerned in the Akassa attack now made submission, but some fled.

Roll No. ADM 171/46 refers to the expedition against Nimbi, Great and Little Fish Town and Tivou Town. Clasps were awarded to H.M. ships Barrosa, St. George, Thrush and Widgeon.

Approximately 360 R.N. personnel and 34 kroomen were entitled to the Clasp.

Roll No. W.O. 100/76 shows 16 officers and 7 civilians were also eligible for the Clasp.

The total entitlement was approximately 417.

BENIN 1897

6TH FEBRUARY—7TH AUGUST 1897

London Gazette No. 26958 of April 19th 1898 gave notice that the medals and Clasps awarded for the Benin Expedition were ready for issue. The grant included the officers, seamen and marines of the undermentioned ships who remained on board at the base during the period as well as those who were actually engaged in the Expedition: Alecto, Barrosa, Forte, Magpie, St. George, Philomel, Phœbe, Theseus and Widgeon.

Apart from capturing Nanna's treasure, the 1894 naval expedition was not particularly successful and the Consul, Mr. J. R. Philips, had unavailingly asked for reinforcements to deal with the situation.

Chief Overiami who ruled in Benin City ignored the Government's Order to stop human sacrifices and slave trading and carried on in the time-hallowed 'bad old ways.'

Ultimately, in spite of being refused permission to visit, Philips, then Acting Consul-General, determined to set off to Benin with a small escort of Constabulary. In January, 1897 his unarmed party was attacked by Overiami's forces about 12 miles from Benin City and all except Captain Alan Boisragon and Mr. Locke, District Commissioner of Warri, were killed. These two suffered great hardships trekking back through the dense bush to the coast.

In the robust words of a contemporary historian, 'Immediate reprisals were necessary, both to avenge the massacre and also prevent the news of an unpunished aggression spreading to the surrounding country and causing revolts and other violent actions in adjacent districts.

Once again, the Navy answered the call with Rear Admiral Rawson in command. Admiral Rawson's despatch was published in the London Gazette of 1897 on page 2516.

The main base was at Warrigi on the Benin River where all available Constabulary and carriers were collected and organised.

The Naval Force consisted of H.M. ships St. George (Capt. G. le C. Egerton)—520 men, Theseus (Capt. C. Campbell)—544 men, Forte (Capt. R. F. O. Foote)—320 men, Philomel (Capt. M. P. O'Callaghan)—190 men, Phœbe (Capt. T. MacGill)—190 men, Widgeon (Lt. E. D. Hunt)—74 men, Alecto (Capt. L. E. Pritchard)—68 men, Barrosa (Comdr. J. Startin)—159 men, Magpie (Lt. H. V. W. Elliott)—74 men. 1,200 all ranks made up the landing party.

Lt. Col. Bruce Hamilton commanded a force of 500 Niger Coast Constabulary.

So as to avoid exposing the sailors and marines to the deadly effect of the climate, the force was only landed when all was ready.

The advance on Benin was made in three columns and began on the 11th February. The columns nearly ran out of water for at Agagi, 12 miles from Benin, the wells were dry. A flying column was therefore sent forward carrying three days' supply of water, rations and ammunition.

Benin City was reached and captured on the 18th February after a week's fighting as each of the column's were attacked at one time or another. Chief Overiami, his generals and principal Juju priests however escaped.

There had been many human sacrifices with a view to delaying the advance and the British Forces were very shocked by the state in which they found Benin City. Seven pits forty feet deep with up to 15 bodies in each were found, the dead being intermingled with the living. In that climate the effect must have been very unpleasant.

Benin was accidently fired and the whole city destroyed.

The white troops were immediately sent back to the ships and the Constabulary left in charge.

There then was a counter palaver, 'country custom being very strict in demanding an eye for an eye and a tooth for a tooth.' It was imperative that a most severe lesson be given to the Kings, Chiefs and Juju Men of all surrounding countries that white men cannot be killed with impunity and that human sacrifices with the oppression of the weak and poor must cease.

The Naval Force re-embarked after eighteen days ashore.

Casualties were 3 officers, including the Commander of the *Alecto*, and eight men killed, including two who were drowned, whilst a further 3 officers and 44 men were wounded. One officer and 4 men died from the effects of the climate. The constabulary lost 40 killed and 23 wounded.

No fewer than 2,290 fever cases were attributed to the expedition. In fact, most of those who participated went down with fever at least twice.

Rewards included K.C.B. for Rear-Admiral Rawson, three C.B.s, two C.M.G.s and four D.S.O.s.

File ADM 171/46 indicates some 2,500 R.N. personnel were entitled. In addition there were 6 Army officers and 3 naval nurses, and of course some 500 Niger Coast Constabulary, also entitled.

It is one of the few of the African naval campaigns where those remaining on board were entitled to the medal.

1899

Army Order No. 253 of 1900 authorised the West Africa Medal with clasp '1899' to the Imperial and Colonial Forces in the expedition known as 'Bula 1899.'

Army Order 253 also granted the medal with clasp inscribed '1899' to the Imperial and Colonial Forces in the Central Division Expedition in February and March 1899 in the Benin Territories expedition in April and May 1899.

Captain (Brevet Major) C. H. P. Carter led an expedition against Olog Bosheri of the Benins who had been responsible for the murder of Philips and his escort in 1897 and then escaped from Benin City later in the year.

Carter destroyed their war camps and captured Olog Bosheri who was tried and hanged.

Casualties were light as the Benins' gunpowder was damp. Constabulary lost 9 men killed and wounded including 2 officers.

Carter was awarded the C.M.G.

S. NIGERIA

Army Order No. 133 of 1902 authorised the clasp 'S. Nigeria' to all officers and men who took part in the operations in the Ishan and Ulia countries under command of Captain (local major) W. C. G. Heneker, Connaught Rangers, in March, April and May 1901.

This expedition, under command of Major Heneker, consisted of 255 officers and men of the Southern Nigerian Regiment and was undertaken against unfriendly tribes to the north-east of Benin City, called the Ishan Ulia.

Very severe fighting occurred between the 9th March and the 8th May and casualties were 3 killed and 22 wounded.

W.O. 100/393 shows the following entitlement:—

Major Heneker, 15 officers and B.N.C.O.s	16
Southern Nigerian Regiment	286
Total	302

ARO 1901-1902

Army Order No. 234 of 1902 approved the grant of the clasp 'Aro 1901-1902' for Imperial and Colonial Forces employed who took part in the operations against the Aro Tribe under Lieut-Colonel A. F. Montanaro, Royal Garrison Artillery, between the 15th November, 1901 and the 23rd March 1902, both dates inclusive.

This comparatively large operation took place against the Aro section of the Ibo Tribe whose country was between the Niger and Cross Rivers. The area involved was some 120 miles east to west and 90 miles north to south.

The objects of the expedition were:—

1. To open out the Ibo country to peaceful trading and prevent further massacres.
2. Stop the slave trade.
3. Suppress the Long Juju fetish. The Long Juju was a type of oracle concealed in a deep ravine with steep sides in dense forest. No disputes could be settled without judgement from the Juju's priests who pronounced doom on their victims from caves in the sides of the ravine. The juju's judgement was paid for in slaves or wealth and more often than not the unsuccessful litigant was sacrificed and his mutilated corpse piled onto a sacrificial altar. Not only the Ibo complained about this fetish, but neighbouring tribes as well.

The Field Force, brigaded for the first time as W.A.F.F., consisted of 1,150 all ranks S.N.R., 300 all ranks 1 N.N.R. and 300 all ranks Lagos Bn. In all, these comprised 11 companies of infantry with four 75 mm. guns. There was also a special band of scouts and a naval brigade of 80 odd under Lt. Comdr. D'Oyly (H.M.S. Thrush). This vessel was described as a steam launch in the Medal Roll which was signed by Trenchard. Finally, 2,300 carriers were required to carry the baggage, ammunition and supplies.

The Force was divided into four columns, commanded by Major Festing, D.S.O. (R.I.R.), Major Mackenzie (R.A.), Major Venour (Hampshire Regiment) and Major W. G. C. Heneker, and the plan was to firstly capture Arochuku, then mop up around Arochuko, and finally converge on and capture Bende.

The columns advanced by different routes and there was no fighting until the 2nd December when Festing met opposition from stockaded towns and entrenchments and lost 2 officers and 40 wounded. By the 24th December, after further fighting, Arochuku was occupied. the Long Juju was blown up and the nearby war camp destroyed.

The next two months were occupied by patrolling throughout the area, during which time the forces met with varying degrees of opposition.

The columns converged on Bende on the 2nd March. Heneker captured two of the Aro Chiefs responsible for the massacres who were then tried and executed.

H.M.S. Thrush covered the landing of a force at Sabagrega on the South Niger.

That there was considerable fighting in this campaign, mostly in thick bush country, is demonstrated by the casualties, which were 13 Europeans wounded and 27 native troops killed and 140 wounded, whilst a further 70 died of disease.

Honours awarded included a C.M.G. to Major Festing, D.S.O., also eleven D.S.O.s and six D.C.M.s.

File ADM 171/56 at the P.R.O. shows 54 of the crew of H.M.S. Thrush entitled, though eight of these medals were returned to the Mint in 1922.

File W.O. 100/390 shows the following entitlement:—

Misc. including a nurse	3
Officers and B.N.C.O.s	39
Lagos Bn. W.A.F.F.	237
Battery N.N.R. of W.W.F. all ranks	15
1 Bn. N.N.R.	145
2 Bn. N.N.R.	157
S.N.R.	1,140
S. Nigerian Civil Service	94
S.N. Gunboat Jackdaw	29
Total	1,909

The Jackdaw was a river gunboat of 85 tons, 100′ x 20′ armed with 2 6-pdrs. Built by Yarrow in 1898, it was transferred to the West African colonial authorities in December, 1898.

S. NIGERIA 1902

Army Order No. 49 of 1906 authorised the clasp 'S. Nigeria 1902' to all officers and men who took part in the following expeditions:—
(a) Under the command of Capt. P. K. Carre (now deceased), Royal Warwickshire Regiment, in the Ngor country in July, 1902.
(b) Under the command of Capt. A. J. Campbell, D.S.O., 19th Hussars, in the Ebeku country in September, 1902.
(c) Under command of Capt. A. D. G. Grayson, Royal Artillery, in the Ikwe country in October, 1902.
(d) Under command of Brevet Lieut-Colonel W. C. G. Heneker, D.S.O., Connaught Rangers, in the Ibeku Olokoro country between the 26th October and 8th December, 1902, both dates inclusive.
(e) Under command of Captain and Brevet-Major I. G. Hogg, D.S.O., 4th Hussars, in the Ibekwe country in October, 1902.
(f) Under command of Captain and Brevet-Major W. J. Venour, D.S.O., Royal Dublin Fusiliers, in the Nsit country in December, 1902.
(g) Under command of Captain and Brevet-Major H. C. Moorhouse, Royal Artillery, in Asaba Hinterland in December, 1902.

W.O. 100/393 shows a total of 51 officers, B.N.C.O.s, and civilians entitled, and 508 ranks of the Southern Nigerian Regiment, a total of 559 clasps.

It will be noticed that the clasp 'S. Nigeria 1902' was not authorised until 1906, a year after the authorisation for '1902–03.'

S. NIGERIA 1902-03

Army Order No. 4 of 1905 awarded the clasp 'S. Nigeria 1902-03' to all officers and men who between the 7th July, 1902 and June 1903, both dates inclusive, took part in the operations in Southern Nigeria under Colonel A. F. Montanaro, C.B., Royal Artillery, against the Uris and the people of Omonoha and Ebima respectively and the operations under Lieut-Colonel W. C. G. Heneker, D.S.O., Connaught Rangers, against Chief Adukukaiku of Igarra.

These operations under command of Col. Montanaro took place in the Oguta area on the east bank of the Niger.

The reasons for the expedition were to stop slave raiding, the attacking of convoys and also to destroy the Igwe Juju which was terrorizing the area.

Col. Montanaro mustered 350 rifles at Oguta on the 3rd April and there was fighting at Umbidi, Amokya and Omachima, where Chief Islobi surrendered.

It took four hours to capture an entrenched camp at Omonoha and thereafter the various towns which had collaborated in the disorders were punished and the Igwe Juju destroyed.

Finally, in May, the town of Elima was taken.

Casualties were 6 Europeans wounded, 13 other ranks killed and 65 wounded, while the porters suffered 22 casualties.

Montanaro was awarded the C.B. Two D.S.O.s and two D.C.M.s were also awarded.

Chief Adukukaiko of Igarra was a bandit who was raiding and generally causing trouble in his district between the Niger and Anambra Rivers.

On the 24th February Major Heneker's force, consisting of 200 rifles of S.N.R., attacked the stockaded town of Alede which was captured after strong resistance and a turning movement executed by Lt. Ward.

British casualties were 1 man wounded.

Army Order No. 49 of 1906 extended the award of the clasp 'S. Nigeria 1902-03' to all officers and men under command of Brevet-Lieut-Colonel W. C. G. Heneker, D.S.O., Connaught Rangers, who took part in the expedition in the Afikpo district in December 1902 and January 1902.

Colonel Haywood in his 'History of the R.W.A.F.F.' says there are no details of the Afikpo campaign available.

File W.O. 160/393 shows 576 entitled to this clasp, mainly from the Southern Nigerian Regiment. Some six civilians and four doctors were also awarded the clasp.

S. NIGERIA 1903

Army Orders numbered 1 and 49 of 1906 awarded the medal with clasp 'S. Nigeria 1903' to all officers and men who took part in the following expeditions:—
A.O. 1/1906
(a) Under command of Brevet-Colonel A. F. Montanaro, C.B., Royal Artillery, on the Nun River from September to the 1st October, 1903.

(b) Under command of Brevet-Major A. M. N. Mackenzie, Royal Artillery, in the Eket district between 16th and 25th September, 1903 both dates inclusive.

(c) Under command of Brevet-Major A. M. N. Mackenzie, Royal Artillery, in the Mkpani country from the 1st to 5th December, 1903, both dates inclusive.

A.O. 49/1906

(a) Under the command of Captain, H. M. Sproule, 4th Cavalry in the Ebegga country in February, 1903.

(b) Under the command of Captain D. L. Roddy, Cheshire Regiment, in the country west of Anan, in March, 1903.

NUN RIVER

This expedition was against a pirate chief called Bibikala who had a stronghold on Wilberforce Island on the Nun River portion of the Niger Delta. His depredations had terrorised the surrounding countryside.

Colonel Montanaro's force, consisting of 326 all ranks S.N.R., a 2.95″ gun, 6 maxims and a gardner gun, assembled at Akassa on the coast and embarked on a number of river craft. 9 Royal Marine officers, a political and a medical officer accompanied the flotilla.

They hunted Bibikala from creek to creek and eventually caught him on the 1st October. He was executed by hanging.

There were no British casualties.

Montanaro received the C.B.

EKET

This expedition in the Eket District was to punish those persons involved in the attempted murder of Assistant Commissioner E. C. Crewe-Read.

Major Mackenzie's column consisted of 6 Europeans, 185 Infantry S.N.R. and 2 maxims. He was accompanied by a doctor and a political officer.

Capt. Roddy, O.C. of the advance guard, was attacked, but repulsed the attacks and captured the towns of Efoi and Ekpa, which surrendered on the 20th and 25th September respectively.

MPKANI

This expedition was against Chief Mpkani who had failed to keep his promise to keep open roads in his tribal area and to stop cannibalism.

Major Mackenzie concentrated a force of 13 Europeans and 288 all ranks S.N.R. at Ugep on the Calibar River on the 1st December.

The column was attacked as soon as it entered Mkpani's country. There was some fighting and the casualties were 3 sergeants wounded, 2 British and 1 African.

Chief Mkpani submitted on the 5th December.

EBEGGA AND ANAN

No details of these patrols have been found, which were believed to be for the suppression of secret societies, cannibalism and general defiance of authority.

A total of 751 persons were entitled to this clasp, apart from 18 others, doctors and civilians in the political service. All the recipients belonged to the Southern Nigerian Regiment, and File W.O. 100/343 refers.

S. NIGERIA 1903-04

Army Order No. 1 of 1906 awarded the clasp 'S. Nigerian 1903-04' to all officers and men under command of Brevet-Major I. G. Hogg, D.S.O., 4th Hussars, who took part in the expedition against the towns of Osea, Oriri, Ndoto between 24th December, 1903 and 15th January, 1904 both dates inclusive.

331 members of the Southern Nigerian Regiment were eligible to receive this clasp (File W.O. 100/393).

S. NIGERIA 1904

Army Order No. 1 of 1906 awarded the clasp 'S. Nigeria 1904' to all officers and men:—
(a) Under the command of Brevet-Colonel A. F. Montanaro, C.B., Royal Artillery, and Captain H. C. MacDonald, Argyll and Sutherland Highlanders, in the Northern Ibibio District between 12th January and the 31st March, 1904, both dates inclusive.
(b) Under the command of Brevet-Major I. G. Hogg, D.S.O., 4th Hussars, against the natives of Asaba Hinterland between the 17th January and 25th April 1904, both dates inclusive.
(c) Under the command of Brevet-Major I. G. Hogg, D.S.O., 4th Hussars, in the Kwale country between the dates of 21st March and 24th April, both dates inclusive.
(d) Under the command of Brevet-Major H. M. Trenchard, Royal Scots Fusiliers, in the Owerri District on the right bank of the Imo River, in March, 1904.
(e) Under the command of Captain H. H. Sproule, 4th Cavalry, and Lieutenant R. D. Whigham, Lancashire Fusiliers, at Obokum and

the patrol under Brevet-Major H. M. Trenchard, Royal Scots Fusiliers, between 3rd February and 3rd June, 1904, both dates inclusive.

As will be noted, the dates for Major Hogg's patrols into Asaba Hinterland and Kwale country overlap.

Northern Ibibio

The object of this expedition was to subdue the hostile North Ibibio District situated on the right bank of the Cross River.

Colonel Montanaro's force comprised 15 officers, 5 B.N.C.O.s, 425 rank and file S.N.R. and two 2.95″ guns. This force concentrated at Mibakpan on the 16th January, 1904.

Two columns traversed the district and in places considerable opposition was met, but by the end of March all towns in the area had been disarmed.

Casualties were one private wounded and one missing, who was subsequently found half eaten.

Asaba Hinterland

The members of a secret society called Ekumeku (The Silent Ones) had instigated a rising by murdering friendly natives and destroying mission stations.

Capt. Hogg's force, consisting of 215 rank and rile of the S.N.R. with a 7-pdr. gun, left Asaba on the 17th January, 1904 and had a battle at Ukungzu where one of the British N.C.O.s was killed.

He was reinforced in February by Capt. H. P. Gordon (Connaught Rangers) with 3 Europeans and 90 other ranks with a 2.95″ gun.

The natives stood and fought again at Okuruku where they were routed and the place destroyed.

300 of the Ekumeku Society were captured and handed over to the civil power for trial.

Kwale

The Kwale people had been at constant war with their neighbours.

Capt. Hogg's column, consisting of 288 men, one 2.95″ gun and two maxims, set out to patrol the Kwale country. A battle took place at Atua on the 26th March where the opposition was crushed and the six chiefs responsible for the tribal warfare and human sacrifices surrendered.

By the 24th April commerce and trade were again in full swing.

OWERRI

The object of this expedition was to pacify the Owerri District on the right bank of the Imo River.

Major Trenchard's force consisted of 4 officers, 2 B.N.C.O.s, 211 ranks S.N.R. and two maxims and it carried out a series of night operations which pacified the district.

Major Trenchard (later first Marshal of the R.A.F., Viscount Trenchard, G.C.B., O.M., G.C.V.O. & D.S.O.) was posted to Calabar as Second in Command S.N.R. in 1904 with the promise from General Kimball that he would have the right to lead all expeditions. He did not see eye to eye with Colonel Montanaro, the Commandant, over laxity in the mess and when the regiment left on an expedition. Trenchard was left behind at Owerri with 40 men as Montanaro claimed he was not acclimatised. He appealed over his head and was ordered to supersede Montanaro, which he did in the field. It seems therefore, that for part of March, Colonel Montanaro was not in command of the Northern Ibibio expedition.

Trenchard's biographer describes how on patrol he came upon a ritual murder. A naked woman strung up from the branch of a tree, head downwards, mouth and private parts coated with honey, to be eaten alive by ants. Her crime was to have been the mother of twins.

There was a skirmish in which five or six men were slightly wounded and thirty Africans killed. The Chiefs then surrendered and handed in their guns.

Trenchard's policy was to spare all but the most stubborn villages and to treat the tribesmen fairly and bluntly. He pointed out, for instance, that having confiscated their guns it was a waste of time for them to buy more as he would merely come back again and remove them.

OBOKUM

Rebels in the German Cameroons attacked and looted Nsana Kang close to the British frontier town of Obokum and slaughtered the German inhabitants. The German rebels overflowed into British territory and persuaded the local inhabitants to attack Obokum between the 10th and 28th of March.

Capt. Sproule's force, comprising 267 rank and file S.N.R., concentrated at Okuni and in cooperation with Colonel Muller, German Army, patrolled the Upper Cross River area and pacified the country without firing a shot.

The German could not understand the peaceful British way of achieving pacification as he believed in punishing tribesmen who defied the law.

Two D.C.M.s were awarded for these actions.

A total of 910 clasps were awarded for the Nigerian campaigns in 1904, of which all went to the Southern Nigerian Regiment.

S. NIGERIA 1904-05

Army Order No. 128 of 1906 approved the award of the clasp 'S. Nigeria 1904-05' to the military forces under command of Brevet-Major H. M. Trenchard, Royal Scots Fusiliers, who patrolled through the unsettled portion of the Ibibio and Kwa country between the 15th November, 1904 and 27th February, 1905.

Again Trenchard used night patrols.

No information regarding this expedition is available except that Trenchard had cleaned up the S.N.R. by prohibiting gambling, womanising, slovenliness and drunkenness.

Col. Montanaro, Major Trenchard and 17 officers and B.N.C.O.s were entitled to this clasp together with 9 civilians and 360 other ranks of the S.N.R. These medal rolls are signed by Trenchard.

Officers and other British entitled:—

Bt. Col. A. F. Montanaro, C.B.	Col. Sjt. W. T. Lindley
Bt. Major H. M. Trenchard	Sjt. J. M. Bradley
Capt. H. C. MacDonald	Sjt. J. Moore
Capt. E. de H. Smith	Mr. C. R. Palmer
Capt. W. J. S. Hosey	Mr. Maxwell
Capt. G. N. Sheffield	Mr. L. T. Marchant
Capt. W. D. Bryne	Mr. H. B. Mansherd
Lt. E. A. Steel	Mr. P. H. A. Grant
Lt. J. F. Mackay V.C.	Dr. T. B. Moffat
Lt. A. A. C. FitzClarence	Dr. J. R. Gordon White
Lt. R. L. Lloyd	Engineer H. Fitzpatrick
Lt. G. S. C. Adams	Engineer N. B. Brooks, Political Dept.
Lt. R. D. Whighan	
Lt. N. M. Geep	C. Police Mr. Richard Marshaw
Col. Sjt. C. Ballersby	Mr. Eyo Hogan Bassey
Col. Sjt. W. W. Pritchard	

S. NIGERIA 1905

Army Order No. 277 of 1906 authorised the Clasp 'S. Nigeria 1905' to all officers and men belonging to the forces employed in the Kwale District between 10th and 18th October, 1905, both dates inclusive.

There had been more unrest in the Kwale District following Major Hogg's expedition of the previous year and the District Commissioner, Mr. Davidson, therefore decided to tour the district. His escort was commanded by Captain Vassal and consisted of Lt. Irvine, 1 B.N.C.O. and 70 other ranks of the S.N.R.

There was a skirmish at Ijonnema on the 12th October where the enemy fought desperately and the force suffered a reverse and had to retire. The D.C. and both British officers were wounded, one soldier was killed and 21 wounded.

Bt. Major Maclear, with 246 men from the Lagos Bn. and one 2.95″ gun, were dispatched to the area and they patrolled the country meeting little opposition and arrested the Chief responsible for the unrest.

It seems that this clasp was awarded only to those who did not take part in the 1905-06 campaign, for which no dates are specified. Some 82 clasps in all were awarded to the Southern Nigerian Regiment, of which 38 were for the medal and clasp. This medal with the single clasp must therefore be the rarest in the whole series.

Officers and NCO entitlements:—

> Capt. P. S. Vassall
> Lt. C. A. L. Irvine
> Col.Sjt. E. A. J. Mayne
> Mr. J. Davidson
> Capt. M. R. Elliott Snr.
> Mr. F. S. James Cm.G. Political
> *Lt. H. A. Child, RN S.N. Marine
> *Lt. H. C. B. Cheetham R.N.B. S.N. Marine
> Dr. A. N. S. Smythe Medic. Dept.
> Native Chief Dore
> Cpl. Oviopo Police
>
> *Clasp only

Total 82, approx. 38 medal and clasp.

S. NIGERIA 1905-06

Army Order No. 277 of 1906 authorised the Clasp 'S. Nigeria 1905-06' to all officers and men belonging to the columns which concentrated at Bende and Oka under the commands of Brevet-Major J. M. Trenchard, Royal Scots Fusiliers, and Captain G. T. Mair, Royal Field Artillery, and took part in the Bende—Onitsha Hinterland expedition.

It should be noted that this is the only Army Order authorizing a clasp which does not give the date of the expedition.

The object of this expedition was to pacify and control an area lying to the south of latitude 6° 30′ N bounded on the west by the Oka Oguta road, on the east by the line joining Afikpo and Ababkaliki, and on the south by the Bende Owerri road, because human sacrifices and slave trading had stopped trade. Even as late as 1905 much of this area had still not been visited by the Administration.

The casus belli was the murder of Dr. Stewart in the Owerri District. This medical officer was new to Nigeria and was travelling alone through the bush on his bicycle and lost his way.

Accounts vary as to what happened. Some say Stewart was murdered as he slept, others that he was captured, trussed up naked and carried alive from village to village with bits being cut off as 'juju' until finally he was decapitated and eaten. All however agree that his body was cut up into small pieces and distributed around for consumption as a fetish. All those who ate a piece of Dr. Stewart would henceforth be released from the white man's domination and protected from any harm from a European.

Thus armed, the whole district rose in revolt.

Major Trenchard's force consisted of the Bende column of 10 Europeans, 325 rank and file S.N.R. and two 2.95″ guns under his immediate command; and the Oka column, commanded by Captain G. T. Mair, R.F.A., consisting of 8 Europeans, 200 rank and file S.N.R. and two maxim guns.

The two columns joined forces on the 30th November and a base camp was established on the Imo River.

There was severe fighting which did not really end until those responsible for Stewart's murder were captured, summarily tried and hanged. The campaign in fact continued until the 15th April by which time 1,100 miles of country had been covered.

Dr. Stewart's skull and bones, except for the hands and left leg, were eventually surrendered and sent to Calabar for burial.

3,000 guns were also surrendered during the operation, which was one of the biggest to take place in South Nigeria.

Casualties were 5 officers wounded, one soldier killed and 58 wounded.

Trenchard and Mair received the D.S.O.

File W.O. 100/393 shows that 48 officers, B.N.C.O.s, doctors and political officers were eligible for this clasp and 657 other ranks of the Southern Nigerian Regiment.

WEST AFRICA 1906

Army Order No. 102 of 1908 awarded the Clasp 'West Africa 1906' to all officers and men under the command of Capt. W. C. E. Rudkin, D.S.O., Royal Field Artillery, composing the Owa columns which left Asaba on the 9th June 1906 and operated under his orders against the people of Owa until the restoration of peace and the breaking up of the columns on the 3rd August 1906. Also to all reinforcements who reached that officer, and operated under his orders against the people of Owa.

This action took place in Southern Nigeria, and the Owa Expedition saw some very heavy bush fighting.

On the 9th June 1906, Mr. O. S. Crewe-Read, Assistant District Commissioner of the Asaba District was murdered at Owa. near Agbor, which is west of the Niger River.

The Ekumeku (The Silent Ones) again began giving trouble after Captain Hogg's 1904 expedition and in June 1906 they surrounded Crewe-Read's camp in the bush. His police escort realised what was happening and they all escaped from the camp in the dark. Crewe-Read, realising his dog had been left behind, went back to fetch it but was set on and killed.

Most of the police escort managed to reach Asaba where Captain Rudkin was stationed with a company of S.N.R. and he immediately went to the place where the murder had occurred.

The whole district was up in arms and the Owa, who were well armed with Dane guns, were operating from thick bush and were hard to locate.

There was very severe fighting around Agbor on the 11th June where the rebels had gathered in great force and this determined resistance continued into July. By this time half of Rudkin's men were casualties.

He was reinforced by a second company of S.N.R. and one 2.95" gun on the 19th June and on the 2nd July a further 3 officers and 130 other ranks arrived.

By now the rebels had retreated to Owa which was assaulted and captured. This broke the main rising although some towns did not submit until August.

This expedition was involved in some of the heaviest bush fighting and casualties were high with 3 British Officers wounded and 12 other ranks killed and 193 wounded.

File No. W.O. 100/394 shows the following entitlement to this Clasp:

Mounted Infantry N.N.R.	47
2 N.N.R.	155
Southern Nigerian Regiment	411

N.N. Police	6
Political officers, etc.	8
	627

WEST AFRICA 1908

Army Order No. 160 of 1910 authorized the Clasp 'West Africa 1908' to all officers, warrant officers, non-commissioned officers and men of the forces under the command of Lieut-Colonel G. F. A. Whitlock, Royal Engineers, Chief Commissioner Anglo-German Boundary Commission; Captain E. G. Heathcote, The King's Own (Yorkshire Light Infantry), Commanding Anglo-German Boundary Commission escort and Lieutenant (now Captain) H. L. Homan, Middlesex Regiment, respectively, who took part in the fighting near the Sonkwala Valley on 11th December 1908 and between the 24th and 31st December 1908, both dates inclusive.

The Yola Cross River Boundary Commission was a remarkable example of co-operation between the British and Germans and I believe one of the few, if not only, occasions when a formation of German troops received a British campaign medal.

This brief account of what happened is based on the official British and German Gazettes and publications.

What is unusual is that the 'Forces under Command' included Oberleutnant von Stephani, Feldwebel Schultz, Vize Feldwebel Bucholz and 40 riflemen of the Schutztruppe fur Kamerun, all of whom received the medal and clasp and, some, the German Emperor's permission to wear.

The Africans involved, or 'Bushmen' as they were then known, were from the Gala tribe from Maitumba who lived in the Sonkwala Valley, Nigeria, between Ikom and Obupa near the border with the German Cameroons, some 150 miles from the Coast.

Some of the Africans in this area had never seen a white man and others had only had hostile contact with Europeans, but they were well armed with dane guns and proficient in staking roads and digging pits with spears in them to catch the unwary attacker.

The British Despatches were published in the London Gazette of 21 December 1909, whilst the Germans published their account in the Deutsches Kolonialblatt of 1909 with further references to personnel in the 1910 and 1911 Kolonialblatts.

Between November 1907 and April 1909 a joint Anglo-German Boundary Commission known as 'The Yola Cross River Boundary Commission' surveyed the Boundary between Southern Nigerian and the German Cameroons. Lt.-Col. G. F. A. Whitlock, R.E., was the British Chief Commissioner and Oberleutnant von Stephani the German Boundary Commissioner. The modus operandi was for the British and German teams to conduct their surveys within their own territories so that the maximum land area on both sides of the Boundary Azimuth could be covered and the results of both surveys would later be compared.

The climate with long periods of continuous rain which transformed the plains into swamps, later followed by the Harmattan wind which prevailed during the whole of the dry season, made life uncomfortable. The terrain in the Cameroon Massif forced the survey parties to cross countless rivers in land without tracks, overgrown with elephant grass and sometimes on slopes so steep that ropes had to be used.

As can be imagined it was difficult adequately to supply the survey teams and their escorts and there were many casualties from sickness. The Germans, to quote the Deutsches Kolonialblatt were proud that the survey was completed in a short time, and, compared to the British Commission, carried out with half the white personnel due to the exemplary fulfilment of their duty by the various Europeans, officers and N.C.O.s, without whom neither the good will of the native soldiers nor the passive obedience of the porters could have been brought together to such successful co-operation.

The Escorts to the survey parties were provided by the 1st Southern Nigerian Rifles under command of Captain C. E. Heathcote (K.O.Y.L.I.) and 91 native troops of the Schutztruppe fur Kamerun (Imperial Cameroon Defence Force) under command of Oberleutnant von Stephani seconded from the German Imperial Army. Armed Escorts were indeed necessary as there had already been an attack on Leutnant von Hanstein at Didan in October 1908, and another attack followed in March 1909 on Leutnant Detzner when he was wounded.

On 10 December Lt. Homan 1st S.N.R. with escort, carriers and loads marched through Chief Maitumbi's country, who had cleared part of the road, without opposition and arrived at Sonkwala where he took over the stores left behind in the care of Chief Bashima. The next morning, accompanied by Colour Sergeant King (Black Watch) 30 rank and file, some carriers and a maxim gun he moved out of camp to reconnoitre to the East side of the Sonkwala valley with a view to selecting a favourable site for a base camp. It was very misty and no opposition was expected until a dane gun went off wounding Pte. Loshunde Ibadan in the arm. The fire was returned, the considerable numbers of Bushmen then made off and Lt. Homan decided to return to camp.

Pte. Alonge Ifon of the rear guard was felled by a single shot and armed Bushmen appeared on all sides in ever increasing numbers, heavy firing took place and the enemy, who several times tried to rush the rear guard, were checked by rifle and machine gun fire. The Galas using parallel roads also tried to cut off parts of the column. The maxim gun was effective in checking the rushes of the enemy, who numbered about 1,000 bushmen, but it went out of action with a broken firing pin.

Colour-Sergeant Phillips back in camp, who could not see what was going on because of the haze, sent out Sgt. Ogo Ibadan and 6 men as reinforcements to the sound of the maxim firing.

A band of enterprising individuals from Chief Ayama's Compound, who were supposed to be friendly, then endeavoured to rush the camp. Colonel Whitlock's carriers, who were saluted by a volley of slugs, fled and the enemy stole the Doctor's boots.

The Bushmen cleared off as soon as the rear party were back in camp, they must have suffered heavy casualties as Lt. Homan could see the strike of bullets and estimated that at least 800 to 1000 of them took part in the fighting, including some of Bashima's men.

On the following day this Chief Bashima offered his assistance with his 'Territorials' saying he was sure there must be a lot of wounded men in the Bush. Lt. Homan fancied Bashima must have got hold of some meat as he had a big dance on that night.

Lt. Homan mentions Colour Sergeant King as being always present wherever the fighting was hottest, steadying his men and encouraging them to fire carefully and himself giving some excellent examples of straight shooting. Of Company Sergeant-Major Bakari Ibadan he wrote 'I do not think I can give him greater praise than to state he was quite as good as a white man in the retirement'; also mentioned were Ptes. Jagun and Ojo Ibadan, all of whom received the West African D.C.M.

The next day, the 12th, apart from the liquidation of two snipers, was quiet and Colour Sergeant Phillips was sent back to Takum with the C.S.M. and 30 rank and file with every available carrier. Lt. Homan kept the maxim and Colour Sergeant King with him for, as he said, he was more than likely to want them.

On 16 December Lt. von Stephani arrived at Takum from Gidan-Sama in order to compare the results of the survey work with Lt. Col. Whitlock and Colour Sergeant Phillips also reached Takum with news of the skirmish.

On 17 December Lt.-Col. Whitlock wrote to Lt. von Stephani 'as I cannot expect reinforcements for at least three weeks, may I ask you if you could possibly send me some of your troops in support as I suppose that you also would wish to see a swift conclusion to events leading to the prevention of a further spreading of hostilities.'

As was reported in Die Festzeitung der Deutschen Herrshalt in Kamerun 'assistance was naturally given immediately and the British

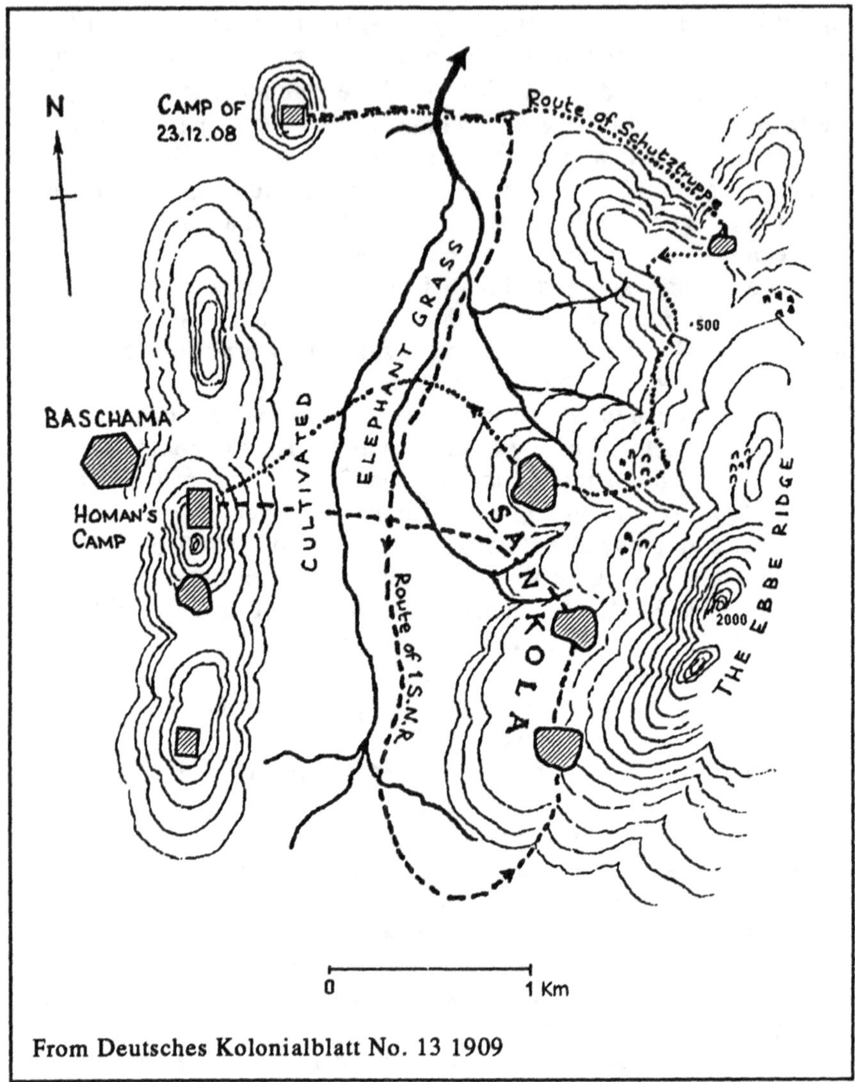

From Deutsches Kolonialblatt No. 13 1909

and German Troops fought together in British Territory. They completely routed the Gayas and put an end to their rebelliousness.'

Lt.-Col. Whitlock with the British Forces under command of Capt. C. E. Heathcote left the next day, picking up the survey escorts at Gashimbela on the 19th and arriving at Sonkwala Valley on the 21st with a maxim, 73 other ranks and proceeded to Lt. Homan's camp, having marched 100 miles in 4 days. All was well at the camp and there had been no more attacks.

Lt. von Stephani's departure was delayed a day waiting for the arrival of his maxim from Gidan Sama but, accompanied by the English HQ staff and joined by the English survey party, he arrived at Sankola (as the Germans spelt Sankwola) on the afternoon of the 23rd, his force consisting of himself, Sgt. Schulz, L/Sgt. Buchholz, 40 troops and a maxim and some 400 porters.

He reported that en route he had had to march defensively as the attitude of the inhabitants was distinctly doubtful and the column was accompanied by hundreds of armed men who continuously surrounded his forces and any lack of purpose must have led to open hostilities.

Capt. Heathcote spent the 23rd in reconnaissance awaiting Lt. von Stephani and issued his orders for attacking the Gaya that evening, the information paragraph of which read:—

'The enemy consisting of the fighting men or the whole of the Sonkwala and their friends from neighbouring valleys are reported to be full of fight. they have blocked and staked all roads leading across the valley from West to East and the roads leading from North to South at the North end of the valley. They have sentries out in the Valley.'

The plan was for the baggage, under command of Lt. Col. Whitlock, to leave the camp at 4.45 am and move to Lt. Homan's camp at Sonkwala. Captain Heathcote and the main body, consisting of Lts. Downes and Hicks, Capt. Beatty the Medical Officer, 2 W.O.s, 49 O.R.s with 20 stretcher bearers and 6 bearers carrying chop and water, to move out at 5 am on the road running East and to attack the enemy from the North. Lt. von Stephani with his force, accompanied by Capt. Moore, was to move out with the main body and attack the enemy near the foothills on the East of the valley. Lt. Homan with 34 rifles of the 1st S.N.R. was to move out of his camp as soon as the baggage party has arrived to co-operate on the flank of the main body. The columns were to mutually support each other.

On the 24th the two columns marched from the Bush camp for half a mile, crossed the river, and separated for their objectives. They immediately lost contact with each other, owing to the dense growth and high elephant grass and in the absence of guides both parties lost their way.

Capt. Heathcote's column marched for 2 hours South along the river, which they crossed twice, and drew near the enemy who could be heard hooting and calling one another. The roads were blocked with trees and prickly stuff with spikes and pits and 2 men had been badly spiked through the foot. At length they emerged into some yam fields, the enemy opened fire and killed a carrier. A fire fight developed the the bushmen would not stand but worked round the flanks. Then several hundred armed bushmen appeared in the yam fields below and the voice of the Chief, prominent in the fight on the 11th, could be heard urging them on to attack. They offered a splendid target for the maxim which unfortunately again jammed and was out of action for some time.

The German column and their maxim could be heard but no very heavy firing was apparent. In the meantime Lt. Homan had also become engaged and his movements relieved the attack on the main column. He joined up with Heathcote and they drove fresh bodies of the enemy into the hills. Pte. Belo Ibadan was wounded and the British returned to the camp at 10 p.m. at night, where they were surprised to hear of the very heavy fighting the Germans had experienced.

What happened to the Schutztruppe is best described in Lt. von Stephani's report in the Deutsches Kolonialblatt:—

'The German detachment set off on the prescribed route ... the beaten track led in a south east direction, round a spur of the Ebbe Range, with the result that communication with 1 S.N.R. was lost from the beginning. I tried to re-establish communication by crossing the spur. Due to the wide expanse of the valley and the irregularity of the country covered with thick bush, it was, however, impossible to regain communication from then on.

'During the march on the heights of this spur, weak enemy forces were driven off by the fire of the advance party. When coming down into the valley I became aware of heavy firing coming from village A. I now followed one of the tracks running along the slope of the Ebbe Range and marched in the direction of the firing with the intention of supporting what I had supposed to be an English assault on village A. Approx. 500 metres above the village lay several separate positions hidden in the undergrowth. From these, the advance party received heavy fire. The advance succeeded in pushing their opponents back and, with the main party and the machine gun, in capturing one of the positions. As the enemy were pressing forward, the rearguard was obliged to fight its way out from this position. During this, Groom Djattau was killed and L/Sgt Buchholz was wounded by a shot in the right wrist. Fighting continued here for a further quarter of an hour.

'Renewed firing from south west of village A reinforced my belief that the English assault on the village was continuing, I therefore gave the order to attack village A in order to hinder the enemy's escape to the hills. Approx 150 metres from the village boundary I deployed the entire detachment in an appropriate position, whilst some patrols took care of those enemy who were pushing from above. After approx. a quarter of an hour of fire, with No. 1 Platoon I took the village outskirts by storm. The machine gun and No. 2 Platoon gave fire support to this assault and followed as soon as the first platoon had opened fire again. When the two platoons joined up, the whole detachment advanced a short distance further. During this advance, I and Soldiers Womma and Ugame, together with Translator Buba, were wounded.

'We proceeded in short steps towards the village but progress was considerably impeded by the profusion of ditches, walls, hedges and undergrowth and was tenaciously defended by the inhabitants. We heard reinforcements arriving from all sides, probably inhabitants of other villages who, having fled from the heavy English column, were now united under a chieftain in attacking the weaker German detachment.

'Having reached a small yam farm outside the village, the detachment was completely surrounded. They were fired on from all sides from a distance of approx 30 metres from positions in the elephant grass which surrounded the farm. I estimated our opponents to be between 600 and 800. A heavy skirmish followed in this yam field during which the machine gun did sterling service. Soldier John Maier was killed, Sgt. Schulze received a flesh wound right upper arm, Soldier Assolo and two machine gun carriers also had superficial wounds. Soldiers Lauani and Efumba II had similar wounds. As after half an hour of fire, the enemy fire diminished, the detachment and I proceeded to break through the enemy position. As there was no track, we had first to make a path through the tall elephant grass. Only odd enemy patrols came after us. Freedman Piu was severely wounded by one of these during a river crossing.

'At 1430, we made a stop downstream from the English camp and the wounded had their wounds dressed for the first time. An hour later, the detachment entered the English camp; I gave charge of the camp to L/Sgt. Buchholz.'

He added:— 'Captain Moore RE stood by me most devotedly after I had been wounded; also, the help given at the camp by the British Expedition's Doctor, Doctor Beatty, to the wounded deserves the highest praise.'

Lt. von Stephani wrote in his initial report, included with Lt. Col. Whitlock's despatches,:—

'I with to mention Capt. Moore especially for his action both before and more especially after my being wounded. During the difficult descent he supported me in person to enable me to reach camp in safety.'

On Christmas Day there was little activity. Assistant District Commissioner Falk arrived at midday from Ikom with 25 rank and file of the 1st S.N.R. and another maxim. The enemy had been very noisy all Christmas Day and night and appeared very anxious to renew hostilities.

On the 26th Capt. Heathcote with 120 other ranks of the 1st S.N.R., a German detachment and 3 maxims, moved out to attack the compound which had given so much trouble to the Schutztruppe. After a little desultory fighting and much hooting the enemy thought better of it and made off; some casualties were inflicted on them with the maxims at medium ranges. The only casualty that day was an askhari badly staked through the foot.

That really settled the matter: the Gaya had had a lot of casualties and the news of the heavy losses sustained by them would not have failed to make an impression on their neighbours. The British losses were 3 killed and 9 wounded and the Germans lost 2 men killed and 9 wounded. 6,770 rounds of ammunition were fired.

Reinforcements of 1½ Companies, 1st S.N.R. arrived in January and Lt. von Stephani returned to Esu on the 10th. Although described as 'dangerously wounded' by Lt.-Col. Whitlock and 'seriously wounded' by himself he does not appear to have required hospitalisation.

The military were notoriously keen to obtain medals for the small wars of the Empire and it is hard not to wonder if there was some self interest when Lt. Col. Whitlock in his despatch to the Governor of Southern Nigeria dated 29th of December 1908 (2 days before the 31st, the final date limit for the clasp) wrote:—

'I trust your Excellency will agree that the operations were carried out most satisfactorily in a difficult country and that all the troops displayed the best qualities of the soldier and I therefore feel I should not be doing justice to the mixed force under my command were I not urgently to put forward their claim to some reward in the shape of a medal, which I hope your Excellency will see your way to recommend. I am certain that the grant of one would be most highly appreciated by Oberleutnant von Stephani and the troops under his command.'

Writing on 12 February 1909 the Acting Governor recommended that a medal be granted in respect of the operations in the Sonkwala Valley which were carried out with skill and ability against strong opposition in difficult country. The Earl of Crewe, K.G., Secretary of State for the Colonies concurred.

Lt.-Col. Whitlock signed the Medal Roll of the Schutztruppen entitled at Chatham on 15 July 1910, which indicates that he had anticipated the issue of the medal to the German forces and Lt. von Stephani had already provided him with a nominal roll of his troops.

The medals were issued by the Foreign Office on 3 January 1911. There can be no doubt that the Germans' medals reached Berlin. It seems probable that the medals were named but this cannot be proved until a specimen is discovered.

The British awards were gazetted on 21 December 1909 with a D.C.M. to Colour Sergeant W. King and the West African Frontier Force Distinguished Conduct Medal for Gallant Conduct at Sonkwala to C.S.M. Bakari Ibadan and Privates Jagun Igberra and Ojo Ibadan.

The German Imperial Gazette of 6 March 1910 shows the award of the 4th Class Royal Order of the Crown with Swords to Lt. von Stephani and that of 6 June to 'former L/Sgt. Bucholz late of the Imperial Defence Force of Cameroon the 'Military Honour Medal, 1st Class''

Later the Imperial Gazette of 21 April 1911 ordered:—

'His Royal and Imperial Majesty has been graciously pleased to give permission to the undernoted Defence Forces of German East Africa to accept the following non-Prussian awards:—

Royal Bavarian Military Service Order 4th Class with Swords Capt. V. Stephani

Royal British Africa Service Medal with Clasp West Africa 1908 Capt. V. Stephani and Sgt. Schulz'

L/Sgt. Bucholz had presumably left the German Army and like the African Askharis of the Schutztruppe apparently did not receive permission to accept the medal or were deemed unworthy of the Emperor's condescension.

It is not possible from the date available to know whether the Askharis' medals were sent to Africa but there is no mention in the Deutsches Kolonialblatt between 1909 and 1915 of this.

Medal Roll W.O. 100/394 records 215 medals or clasps for 'West Africa 1908' were issued to the undernoted British personnel:—

Officers, political officers and N.C.O.s		15
Lt. Col. G. F. A. Whitlock	R.E.	
Capt. C. E. Heathcote	K.O.Y.L.I.	
Capt. M. T. G. Moore	R.E.	
Capt. C. C. G. Ashton	S.N.R.	
Lt. W. V. Nugent	R.A.	
Lt. H. L. Homan	Middlesex Regt.	
Lt. L. D. Hicks	S.N.R.	
2/Lt. W. D. Downes	Royal Sussex Regt.	
Dr. G. Beatty	Medical Dept.	
7285 Cr.Sgt. L. H. Phillips		
6191 Cr.Sgt. W. King	Black Watch	
9870 L/Cpl. M. Locke	R.E.	
L/Cpl. E. F. Davies	R.E.	
Mr. E. Dayrell	Political Dept.	
Mr. E. Falk	Political Dept.	
Askharis of the 1st Southern Nigerian Rifles		157
Kaiserlich Schutztruppe fur Kamerun		
Oberleutnant von Stephani		
Feldwebel Schultz		
Vize Feldwebel Bucholz		
and 40 soldiers		43

It is not a common medal.

Probably many of those issued to the S.N.R. were lost or melted down. In the Great War in spite of overwhelming odds in their favour it took the Allies nineteen months to defeat the German Forces in the Cameroons and it seems unlikely the Schutztruppe would have retained a British medal during such an arduous bush campaign. None is recorded as having come on the market in Germany.

Only Lt. von Stephani's medal may be traceable. His granddaughter handed over all decorations of the family to a lawyer in Cologne in the 1970s, who handed them on to another friend, who gave them to his father-in-law but I have been unable to make contact with the present owner or to find out if his West Africa Medal still exists, or whether in the 'Gott Strafe England' attitude of 1914 it was consigned to the rubbish dump as was the case, I know, with General Thwaite's Prussian 4th Class Order of the Red Eagle.

Hauptman von Stephani was transferred to the Regular Army from the Defence Force with effect from 3 August (Imperial Gazetter 1914 p 772). The 1914 Army List shows Whitlock with the Ordnance Survey in Dublin, having reverted to his substantive rank of Major, Heathcote as a Major superintending Gymnasia in Eastern Command and Captain Homan back with the Middlesex Regiment.

The medal shown in the article is that awarded to Dr. Beatty, it is impressed 'Dr. G. Beatty Medical Dept. SNFF.' He it was who lost his boots and was mentioned in the Kolonialblatt for conduct 'deserving the highest praise.'

Nowadays the 'Bushmen,' who, after all, merely tried to block the roads and paths to their villages and defend their homes, would not be rebellious but freedom fighters trying to repel foreign invaders.

WEST AFRICA 1909-10

Army Order No. 149 of 1911 approved the African General Service Medal (as instituted by his late Majesty King Edward VII), with the Clasp inscribed 'West Africa 1909-10' being granted to all officers, warrant officers, non-commissioned officers and men who took part in the fighting in the Ogwashi-Oku country between the 2nd November and 18th December 1909 and between the 6th January 1910 and 24th April 1910 under the command of Captain (Honorary Major) G. N. Sheffield, 3 Bn. Essex Regiment, and between the 4th and 27th May 1910 under the command of Major G. E. Bruce, Norfolk Regiment, both dates inclusive in each case.

It will be noted that although George V had ascended the throne this medal was issued with Edward VII on the obverse. Also that only one clasp was issued although the hostilities were for quite separate periods in 1909 and in 1910.

Again this trouble was fermented by the Ekumuku Society. They had stopped trade, defied authority and destroyed Government and Roman Catholic mission buildings.

Capt. Sheffield with 4 officers, 1 B.N.C.O. and 131 rank and file with one maxim tried to capture the ringleaders but the bush was very thick and there was strong opposition from the town of Ogwashi-Oku for several days. The ringleaders were not captured and on the 18th December the tribesmen were given three weeks to surrender them. Casualties to date had been one killed and five wounded.

The ultimatum having expired, fighting was resumed and after losing 7 killed and 15 wounded, the column retired to its base and awaited reinforcements.

Major Bruce, with 5 officers and 160 other ranks S.N.R. and two maxims, arrived and operations were resumed in May. Several enemy bush camps were surprised and numbers of tribesmen were killed. Major Chapman (Suffolk Regiment) and 3 soldiers were killed and 3 wounded.

Total casualties were therefore 34, almost 12 percent of the force engaged.

Trenchard, father of the R.A.F., achieved great results by treating his defeated enemies well, and in a conversation, one of the captured leaders said to Trenchard, 'Give me one piss-gun (maxim) and I'll beat you.'

File W.O. 100/394 shows the following entitlement:—

Major Bruce and other officers and B.N.C.O.s	19
Political officers and medical	11
Southern Nigerian Police	12
Southern Nigerian Regiment	385
	427

NIGERIA 1918

Army Order No. 460 of 1924 approved the Clasp 'Nigeria 1918' to the officers and men who took part in the operations against the Egba Tribe in 1918 between the 11th June and 31st July, both dates inclusive, in the vicinity of the Nigerian Government Railway (main line) from Abeokuta in the north to Lagos in the south, within the area bounded on the east by a line from Abeokuta through Ijebu-Ode to Lagos and on the west by a line from Abeokuta to Ilaro and thence through Igbessa to Lagos.

This revolt was known as the Abudi War.

The Egba Kingdom had not been included in the Lagos Protectorate of 1895, but in 1914 the treaty granting the Egba independence was terminated. There was trouble then and troops had to be brought in to quell the Ijemo troubles. His Majesty's Government started to administer this portion of Yorobaland indirectly, more reforms followed and by 1918 the various fiscal measures, including direct taxation, were bitterly resented. There was also a lot of graft by the Native Administration exacerbated by the Ogboni Secret Society.

The British Resident threatened to send for troops, whereupon the Egbas rebelled.

An officer and 50 ranks 4 N.R. and a company 1 N.R. were sent to restore and protect the railway line, and these were reinforced by the 4th Service Bn., Col. Sargent commanding, and 250 from the Ibadan Training Centre.

General fighting occurred in June, mostly in the form of sniping. Capt. Walker attacked and took a rebel camp at Awba, ten miles south of Abeokuta, where a friendly Chief had been murdered, and there was fighting.

Operations were now directed to encircling the insurgents by the converging drives of five columns, all directed on Iraw and Mawkawlawki on the Ogun River south of Awba; No. 1 column (Captain Walker) from the north; No. 2 column (Captain Maxwell) from the centre; No. 3 column (Captain Norman, M.C.) from the south; No. 4 column (Captain Johns) from the east; and No. 5 column (Captain Evans) from Awba and Teppoma; each of two companies.

On the 4th July the movement was successfully completed, most guns were surrendered and the ringleaders taken into custody. Between the 11th and 23rd July, 3,400 guns were surrendered at Lafenwa on the railway.

Our casualties were: W.A.F.F.—8 killed and 67 wounded; police and others—2 killed and 11 wounded. 4 Europeans were also wounded. The fighting had not been severe, although sniping was continuous.

Brigadier Cunliffe, Commanding the Service Brigade, first directed operations. When he and Lieutenant-Colonel Sergeant went on leave Major Hawley took over.

The unbound Medal Rolls comprise 310 loose pages and are larger than usual. There are a large number of entitled personnel who could not be found when the rolls were prepared and a year or so later there are lists of no less than 63 pages of 'untraced men.' Considering that many of the men must have been enlisted for the Kaisar's War only and the medal was only authorised six years after the end of the uprising, when the majority were demobilised, this is not surprising.

It would take a long time to work out precise numbers, but the following units are entitled: The figures in brackets are an extremely rough guide of entitlement.

1st Training Centre W.A. Service Bde.	(324)
1 Bn. N. Regiment	(180)
2 Bn. N. Regiment	(226)
4 Bn. N. Regiment	(259)
4th Training Centre W.A. Service	(21)
H.Q. West Africa Service Bde.	(21)
H.Q. Nigerian Land Contingent Nigerian Police	(89)
H.E. The Governor and Nigerian Civil Service	(7)
West Africa Service Bde. (stokes gun coy)	(26)
4 W.A. Service Bde. (4 N.R.)	(85)
Civilians, Europeans	(37)
West Africa Service Bde. (3 N.R.)	(109)
P. & T., Natives (Nigerian Railways)	(31)

The civilians received their medals for special services and apparently had to have been under fire and the remarks against their names include:—

'rerailing engine derailed by rebels'

'driver: in-charge of armoured train'

'restoration of telegraph, telephone and electrical block signalling destroyed by rebels'

'under fire on break-down train'

It seems probable that about 1,000 medals were granted.

CHAPTER EIGHT

Northern Nigeria

NIGER 1897

Army Order No. 155 of 1897 authorised a medal of the same design as that for operations on the 'West Coast of Africa' with the Clasp inscribed 'Niger 1897' to the forces which took part in the operations to Egbon, Bida and Illorin or formed the garrisons of Fort Goldie and Lokoja between 6th January and 25th February 1897.

This campaign is described in detail in Lt. S. Vandeleur's book 'Campaigning on the Upper Nile and Niger,' and was the first real trial of strength between Sir George Taubman Goldie's Royal Niger Company and the Mohamaden Fulani Tribe.

The Fulani lived in the Niger Valley. They owned slaves, lived in walled cities and fought on horseback. They rode on high peaked saddles with enormous brass and iron stirrups, their ponies were covered with trappings and they were armed with swords and spears. the horsemen were generally dressed in white robes with large turbans and were attended by two or three squires to carry their guns, spears etc. The infantry was armed with guns and rifles.

Bida, the capital of the Nupe State, part of the Hausaland Empire, was a walled town of 6,000 inhabitants situated about 25 miles north of the Niger River. The Emir of Bida had sent a large raiding party across the river and set up a war camp at Kabba which was even threatening the Niger Company's headquarters at Lokoja.

Sir George Goldie determined to put a stop to the Emir's activities, and, incidentally, spread the power of his Company. He obtained officer reinforcements for the Company's Constabulary and concentrated at Lokoja.

Major A. J. Arnold, 3rd Hussars, was in command and had at his disposal about 1,000 Hausa troops of the R.N.C. armed with snider rifles. His force consisted of 30 officers and other Europeans and 513 Constabulary. Their armaments included one 9-pdr. and one 12-pdr. Whitworth field guns, rockets, five 7-pdrs. R.M.L. and six maxim guns of .45″ calibre. There were 900 carriers.

Two Mallams (the West African term for a Mussalman Holyman) accompanied the expedition to prevent the Constabulary being suborned by the Fulanis preaching a jeddah to their co-religionists against the Europeans.

'The guns were dragged by hand at great labour.. the infantry advanced in square with the guns in the centre. A halt was made on commanding ground near the town and the guns were brought into action upon masses of the enemy thronging its outskirts. Thereupon the square was surrounded by hostile swarms who loomed up from all sides, but the steady musketry and maxim fire kept the Fulanis at a distance ....

The tactics were to attack the Emir's army south of the river and prevent the Fulanis joining up by patrolling the river with a flotilla in the charge of William Wallace, Goldie's Assistant Director. This flotilla consisted of two stern wheelers, Empire and Liberty, and twelve other vessels, some of which must have been quite commodious as they could carry up to 300-400 men. Some were armoured with steel plates and carried two Nordenfelt shell guns on wheeled carriages, presumably 1-pdrs.

The force marched on Kabba without contacting the enemy who had retreated north-west. Sir George Goldie there held a ceremonial parade, declared Kabba free and abolished all slave trading.

The force then marched north to Egbon, crossed the Niger, which is 850 yards broad at this point, and continued on towards Bida. At this stage Major Arnold's force consisted of 32 Europeans, 507 Constabulary and 565 carriers, making a total of 1,104.

Their advance was disputed by masses of horsemen estimated at between 20-30,000 in number. The Fulani attacked again and again until they had been driven back to the great ten-foot high crenellated red mud wall of Bida. It was obvious that although the Nupes attacked the square recklessly and with great gallantry they could not succeed against the fire of the artillery, maxims and rifle volleys of the Constabulary who stood firm against the charges of the cavalry. Bida was shelled and the Nupes fled in panic.

The Fulanis' losses were described as enormous, but the British only lost 1 officer and 7 other ranks killed and 9 wounded, which was a small cost for inflicting such a crushing defeat on the Fulani horsemen.

The force, which then re-embarked in the S.W.S. Empire and Liberty and S.S. Soudan and Florence for an expedition against Ilorin, consisted of 21 Europeans, of whom 15 were officers, 320 soldiers, 20 gunners and 488 carriers, a total of 829. Their armament included two 7-pdr. mountain guns and four maxims.

The Constabulary marched south to Ilorin on the 10th February and it was hoped that the town would submit without fighting. However, the column having formed square was charged by Fulani horsemen who showed complete disregard of the British weapons. The square advanced, beating off attacks, and again very heavy casualties were inflicted by the maxims. After two days' fighting the horsemen fled and Ilorin was taken after being bombarded. A treaty was made with Emir Sulimain placing Ilorin under the protection of the Company.

File No. W.O. 100/76 shows the following entitlement:—

Sir George Goldie and 13 civilians	15
British Officers	24
B.N.C.O.s	7
Royal Niger Constabulary	651
European Civilian employees	15
	697

There exists another nominal role of Non commissioned officers at Lokaja and Fort Goldie during the Niger Sudan campaign. This refers to 135 and one British civilian. 3 natives appear on both rolls (reference F.O. 83/1798).

This total is less than the numbers which would appear to have taken part in this campaign.

1898

Army Order No. 253 of 1900 authorised the West African Medal with the clasp '1898' to:—

(a) forces who were employed in the military expeditions known as Ibouza, Anam, Barua, Bassema, Siama, Angiama, Illah and Dama, all in 1898.

(b) to the military forces employed in the Lapai expedition under Lieut-Colonel Pilcher from the 8th June to 27th June 1898.

(c) to the force under Lieut-Colonel J. Willcocks, C.M.G., D.S.O., in Borgu before the 14th June, 1898.

(a) Lt. Keating of the 1 Bn. N.N.R. was sent with a force of 30 men to arrest the Chief of Helo Island in the Illah District. He was ambushed and his party was all killed. Major Morland took out a punitive expedition.

(b) The Emirs of Lapai and Argungu had been slave raiding towns on the banks of the Niger River assisted by the Emir of Bida.

Pilcher set out with 300 asharis from the 1 Bn. N.N.R. and 100 R.N. Constabulary and three maxims. After a forced march they surprised the enemy forces of 600 horses and 3,000 foot and routed them. Their camps and strongholds were destroyed and their power broken. In these actions the troops received their 'baptism by fire' and suffered some casualties.

(c) In 1894 the Royal Niger Company had employed Capt. Lugard to make a treaty with the King of Borgu, granting trading rights, which he achieved in the face of French competition. Lugard's expedition was not without its adventures, as once he felt a blow in his head, tried to take off his helmet, but without success, as it was transfixed by an arrow which had penetrated into his skull.

In 1898 the French were trying to expand from Dahomey and an expeditionary force was sent under Colonel Willcocks with Major Arnold of the R.N.C. The force consisted of 1 and 2 Bn. R.N.C., (with a battery) commanded by Lt-Cols. Pilcher and Fitzgerald respectively.

No fighting took place and the French withdrew. In September 1898 a treaty was signed in Paris which recognised the British claims to Borgu.

1900

Army Order No. 123 of 1903 extended the grant of the West African Medal (A.O.253 of 1900) with clasp '1900' to the Imperial and Colonial Forces employed in the Munshi expedition between 4th January and 19th March 1900 and the Kaduna expedition between the 20th February and 9th May 1900.

These operations were in furtherance of Lugard's scheme for developing the country by ensuring the safe passage for caravans and the abolition of the slave trade.

MUNSHI EXPEDITION

The Munshi Tribe, who lived in the Benue Valley, had been hindering the passage of caravans from Nassarwa and Keffi, but the immediate casus belli was the cutting of telegraph wires.

Capt. Carroll with 1 coy 1 N.N.R. attacked the town of Akwanagion, whose fortifications were a ditch 10 feet wide and 8 feet deep with a stockade. The defenders kept up a heavy fire of arrows and skirmishes continued from the 4th to the 10th January. Akwanagion was then captured and burnt down.

Captain McClintock, R.E., repaired the telegraph line and then reinforced Carroll with 50 rifles.

They then attacked the town of Gidan Barta which in spite of a strong and vigorous defence was captured on the 27th January.

Lt.-Col. Lowry Cole with 300 rifles of 2 N.N.R., three maxims and a section of 7-pdrs. then marched to the northern and eastern frontiers to overawe the rest of the Munshi and subdue any resistance met.

Casualties were 3 natives killed and 15 wounded and included a British N.C.O. who was struck in the foot with a poisoned arrow.

KADUNA EXPEDITION

Col. T. L. N. Morland with a force of 300 W.A.F.F. carried out a reconnaissance to survey a site for Government Headquarters north of the Niger in the area of the Rivers Kaduna, Gurara and Okwa.

The opportunity was taken to attack any slave raiders found, some of whom resisted, particularly at Lima where Colonel Lowry-Cole was wounded by a poisoned arrow.

Total casualties were about 20, including another officer.

Army Order No. 132 of 1902 authorised the striking of a new medal in silver to commemorate military operations in East, Central and West Africa to be designated 'The African General Service Medal.'

N. NIGERIA

Army Order No. 133 authorised the Clasp 'N. Nigeria' to all officers and men engaged in:—
(a) the operations against the forces of Bida and Kontagora and who formed part of the force under command of Major W. H. O'Neill, Royal Artillery, from July to December 1900, or under command of Major (local Colonel) G. V. Kemball, Royal Artillery, between 19th January and the 17th February 1901.
(b) the expedition against the Chief of Tawari under command of Major (temporary Lieut-Colonel) A. W. G. L. Cole, Royal Welsh Fusiliers, between the 6th and 8th December 1900.
(c) the operations against the Emir of Yola under the command of Brevet Lieut-Colonel (local Colonel) T. L. N. Morland, Kings Royal Rifle Corps, in August and September 1901.

BIDA AND KONTAGORA

Major O'Neill a Horse Gunner, with Lt. Porter and a small force of 13 Mounted Infantry and 25 infantry, had many skirmishes with the Nupes as he advanced on Bida, driving large forces of enemy before him and inflicting heavy casualties.

He advanced right up to the walls of Bida, entered the town with a few men and fought the Emir Abudekri in a hand to hand fight. He was badly wounded and only escaped through the timely arrival of his men.

O'Neill had acted against orders in approaching so near Bida, but his duel had made a profound impression on the tribesmen.

Kontagora was built in 1863 in the middle of a highly populated pagan area by Umero, a brother of the Sultan of Sokoto. He and his son, Ibrahim Nagamaste, who succeeded him, devastated the surrounding countryside by wars and slave raiding so that much of this area is still thinly populated today.

Ibrahim was captured in February 1903 and taken in chains to Lokoja, which action hardened the resistance of the Emirs in the north.

Major Kemball was sent out to tackle the Kontagora horsemen who had been attacking our small patrols. His force consisted of 11 officers, 3 B.N.C.O.s and 433 infantry of W.A.F.F. with two guns and three maxims. The force marched northwards along the Niger and then swung to the north to cut off the Emir from his allies. It then marched in square formation against the thousands of infantry and horsemen who gallantly charged the square but were inevitably repulsed with heavy losses.

The town then fell easily and the only British casualty was one man wounded by an arrow.

The old Emir, Sarikin Sudaw, 'The Destroyer,' fled. He was later captured and when reproved for slave raiding he asked his captors, 'Can a cat be stopped from mousing? When I die I will be found with a slave in my mouth.'

Col. Kemball received the D.S.O.

TAWARI

This expedition was to avenge the murder of Hon. D. Carnegie, the perpetrators of which crime came from Tawari. Lieut-Colonel A. W. G. Lowry-Cole with one company of W.A.F.F. made a forced march on Tawari which was strongly fortified and difficult to approach. However, after the gate had been blown open by gunfire, there was little opposition and the ringleaders were captured.

The Fulani Emir of Yola lived in a mud palace beside the Benue River near the eastern border of Nigeria and was anti-British. He had forbidden the Niger Company's officials to land in his Emirate.

His town was the centre of the slave trading in the area and large numbers of pagans were sent from his headquarters to the north as slaves. All efforts to stop him slaving having failed, an expedition was sent.

Colonel Morland and a force of 365 W.A.F.F. with two 75 mm. guns proceeded up the Benue River in sternwheelers to Yola. On arrival there Morland formed square and awaited the Emir's attack. The Emir's forces included 60 deserters from Rabeh's forces from the French area of influence armed with modern rifles. His armament also included two rifled cannons given him by the French, but his main forces were armed with more primitive weapons like bows and arrows.

They attacked Morland's square and after ten minutes fighting fled back to the town which was 1,000 yards or so away. Here, at the

Mosque, Yola's forces made a stand and their two French cannons opened fire and caused casualties. Lt. A. McLintock (Seaforths) and Lt. T. A. Rose (R.S.F.) charged the guns and captured them and our own artillery blew in the palace gates at point blank range. Capt. C. E. G. Mayne (H.L.I.) worked round and cleared out the remaining pockets of resistance and the enemy fled leaving much ammunition behind.

The Yola casualties were high and we suffered casualties totalling 39 and 2 officers, including Colonel Morland, wounded by poisoned arrows.

The Emir escaped and his brother was installed in his place.

Colonel Morland and Major McClintock were awarded D.S.O.s (File W.O. 100/391).

N. NIGERIA 1902

Army Order No. 129 of 1903 authorised the clasp 'N. Nigeria 1902' being granted to the Imperial and Colonial Forces employed in:—
(a) The Bornu expedition between the 1st February and 16th May 1902, both dates inclusive.
(b) The Kontagora expedition between the 12th and 20th February 1902, both dates inclusive.
Army Order No. 4 of 1905 extended the issue of the clasp to officers and men under command of Capt. G. C. Merrick, Royal Artillery, at Arungu and on French Convoy duty between 15th June and 30th November 1902, both dates inclusive.

BORNU

It was decided to carry the flag to the old Kingdom of Bornu and a long circular march was to be undertaken by Col. Morland with 13 officers and 515 troops from 1 and 2 N.N.R., two 75 mm. guns, four maxims and, also 800 carriers. Wallace, the Acting High Commissioner, accompanied the expedition.

They started from Ibi on the Benue River and marched northwards.

The Emir of Bauchi had a large slave market which acted as a distribution point for the slaves from Yola and the pagans surrounding his area. He had recently seized the town of Guaram and massacred or enslaved all its inhabitants.

The Emir fled and was deposed amid general rejoicing and no opposition.

Lt. Col. Beddoes and 1 coy N.N.R. had a skirmish with the Yergums who submitted and Bauchi was then garrisoned and the march continued via Gombe to Gojba in the Bornu district.

Mallam Jibrella, who was a white haired old man with plenty of dash and pluck, tried to emulate the Mahdi of the Soudan and started to lead a Holy War with only 100 horsemen and 600 foot. He attacked but could never get near our lines and he was beaten off by volleys from the infantry kneeling in line and the guns firing case shot. He lost 66 killed and many wounded whilst our casualties were only two wounded.

Mallam fled but was captured after a seventeen hour pursuit. He was then given a small pension and lived for the rest of his days at Korin.

Morland marched on through the black cotton plains of Bornu which had been ravaged by the freebooter Rabeh from French territory. Here he was greeted as a saviour. He returned via Yola to Ibi having brought some 60,000 square miles under administrative control with very little bloodshed.

3 D.C.M.s were awarded.

Against cavalry, square formation was usually adopted. A contemporary parade ground photograph shows a square of 22 ranks of Askaris a side, kneeling with bayonets fixed and with a maxim gun at one corner manned by a European. The company commander and another officer stand in the centre with their chargers held beside them.

H.M.S.O. 'Small Wars' published in 1906 devotes a chapter to squares, stating the raison d'être of the square as 'the necessity imposed on a body of regular troops to being able to show a fighting front in any direction and in the obligation of protecting supplies and wounded.'

KONTAGORA

The deposed Emir Ibrahim of Kontagora collected an army of freebooters and set up as an independent slave trader preying on the pagans in the Zaria area.

A garrison was stationed at Zaria when its Emir requested support against the hostility of the Kontagora freebooters. He had only just accepted British authority when he had to appeal to Lugard for help.

ARUNGU

Capt. G. C. Merrick, Royal Artillery, with an escort of 50 rifles N.N.R. and Lt. A. C. McLaclan (18 Hussars) was sent to ensure the safe passage of French convoys working with the French Boundary Commission on the border between the two territories.

It also had the role of preventing hostilities between Sokoto and Gando.

Apart from one fairly severe skirmish that took place at Gilwazi no fighting occurred. Casualties were one soldier and three followers killed and 3 other ranks wounded.

The books of reference state 525 medals were awarded, but more personnel must have been entitled as this figure does not take into account the Kontagaro and Arungo expeditions. File W.O. 100/391 shows the following listed for this N. Nigeria 1902 clasp:—

Misc. officers and B.N.C.O.s	14
No. 1 & No. 2 Bty. N.N.R.	146
No. 1 Bn. N.N.R.	396
No. 2 Bn. N.N.R.	483
	1,039

N. NIGERIA 1903

Army Order No. 171 of 1903 authorised the clasp 'N. Nigeria 1903' to the Imperial and Colonial Forces employed in the Kano-Sokoto Campaign 1903 between the 29th January and 15th March 1903, both dates inclusive.

Later, Army Order No. 65 of 1905 also authorised this clasp to those officers, soldiers and others of the Imperial and Colonial Forces engaged in the Sokoto Burmi operations between the 15th April and the 27th July 1903.

The casus belli was the murder of Capt. Maloney, the Resident of Keffi. Capt. Maloney sent his assistant, Mr. Webster, to summon the Magaji, who was the representative in Keffi of the Emir of Zaria, to a conference.

Accounts differ as to what happened. Some say Mr. Webster was led to the harem from which he was thrown out for causing disturbance. The Resident, who was still crippled from a wound received some time before and was lying in a litter, was then shot dead by the Magaji with his pistol. Other accounts say that as Webster was well known and a friend of the Magaji, orders were given that he was not to be hurt.

The Magaji then fled on his war horse to Kano where he was honourably received by the Emir.

Kano was a city of between 30–35,000 inhabitants, surrounded by a wall 11 miles long, pierced by 13 gates and was thought to be impregnable. The wall varied from 30′ to 50′ high and was 40′ thick at the base. There was a double ditch in front which was divided by a triangular wall.

When asked to surrender the Magaji, the Emir of Kano sent a reply stating that between him and the British, 'there could be no dealing except as between Mussalmans and Unbelievers, (Kafiri)—War!'

After a series of vacillating cables with Whitehall, Lugard set off for Kano just behind his troops, which as Governor he could not now command.

In January Colonel Morland left Zaria with 24 officers, 13 B.N.C.O.s and 722 rank and file including 101 mounted infantry of the N.N.R., four 75 mm guns and four maxims. The 75 mm guns were adapted for portage, some loads requiring four porters. Captain Abadie, the Resident of Zaria, was appointed political officer, and Captain Porter and Captain Wright accompanied the expedition.

The Colonial Office ordered up reinforcements from Southern Nigeria, Lagos and the Gold Coast, which consisted of 600 rifles S.N.R. and 300 from the Lagos Bn. The whole expedition was placed under command of Brigadier-General G. V. Kemball, Inspector General of W.A.F.F. Colonel Morland was with the leading troops and met the first opposition at Bebji, just inside the Kano frontier. After a parley, a well-placed shell killed the Chief and blew open the door of the town which surrendered.

Kano was reached on the 5th February after encountering no further resistance. Shelling had no effect on the walls even though double common shell was fired, but a breach was eventually made in the Zaria Gate and the Fulani, who numbered 800 horses and 5,000 foot, fled, pursued by shell-fire and the Mounted Infantry. The Fulani suffered a loss of 100. British casualties were 14 wounded, 2 officers and 12 men, including one officer wounded by a sword cut from the Emir's chief eunuch.

The Emir Aliyu's forces had already fled towards Sokoto and were followed by the British Forces. Captain Wright with only 45 M.I. withstood attacks by very large numbers of horse and foot who charged his square ten times. He was awarded the Victoria Cross. There were 65 horsemen lying dead within 30 yards of the square.

The Kano cavalrymen were very impressive. They wore high turbans, flowing robes and chain mail armour and were armed mostly with spears and shields. They were mounted on horses which wore armour of quilted cloth.

Morland's force numbered some 675 all ranks and comprised 25 officers, 5 N.C.O.s, 1 N.C.O. 'medic,' 22 M.O.s 68 gunners 439 infantry 89 M.I., 195 guns and maxim carriers, 400 general duty carriers, 4 maxims and 4 guns.

The forces advanced on Sokoto and a skirmish was fought at Kotorwashi, where once again the disciplined musketry of the W.A.F.F. drove off the attackers.

Outside Sokoto 15,000 horse and 3,000 foot of Fulani had gathered. They did not however stand to fight, but fled after a few shots had been fired. Their casualties numbered hundreds whilst the British had but one man wounded. As one of their chiefs said, 'The guns fired bap-bap-bap and many hundreds were killed.'

The Sultan and the Magaji fled to the borders of Bornu, where the Emir Aliyu was captured by the King of Gobir, his traditional foe, who handed him over to the British. He was later exiled by the British to Lokoja.

In addition to Wright's V.C., the honours awarded included C.B.s for Kemball and Morland, 3 D.S.O.s and 8 D.C.M.s.

Lugard then received a dispatch from His Majesty's Government reproving him for having undertaken these expeditions as only Whitehall had the knowledge 'as to whether it was necessary to send an expedition to Kano.'

Sokoto Burmi

The object of this campaign was to capture Aburahaman, the ex-Sultan of Sokoto. He was at large with a band of fanatical followers, including those of the Magaji of Keffi and Abubekor of Bida, and was considered likely to raise the country against the British.

In April the Resident of Sokoto requested counter-action and the various forces were sent off piecemeal. Captain Goodwin with 60 men of the 1 N.N.R. from Sokoto, Captain Sword with 42 men from Kano, Lt. Grozier with 25 M.I. from Zaria, Major Plummer with 2 officers and 70 O.R.s from Bauchi, Lt. Hon. D. Carleton with another 15 M.I. from Zaria.

Lugard in his dispatch to London gives those engaged as: from Kano—2 B.O.s, 1 B.N.C.O., 64 men and a maxim; from Zaria—4 B.O.s and 60 M.I.; from Bauchi—2 B.O.s, 1 B.N.C.O., 100 men and a maxim.

There were some skirmished and Aburahaman was pursued by Captain Sword along the Gongola River to Burmi. He attacked Burmi, but his assault was repulsed with the loss of Major Plummer and Sergeant Hayes wounded, 2 other ranks killed and 54 wounded, approximately 43 percent of the force. Many of the wounds were from arrows.

The enemy were described as 'lying literally in heaps, and had fought with fanatical bravery.' As Captain Sword's ammunition was running low he withdrew to Bandu. On the 1st June Major Barlow and 100 or so other ranks drove the enemy out of Burmi and killed a large number.

On the 20th June further reinforcements had arrived under Major Marsh who assumed command of a total of 445 infantry, one gun and three maxims.

On the 27th June there was a hard fought battle Burmi, was stormed and Aburahaman and the Magaji of Keffi were killed, together with 600 to 900 of their followers.

The garrison of 10,000 were armed with bows and arrows, throwing spears, Dane guns and rifles.

The bowmen in trenches staunchly held their positions although under heavy shrapnel fire from the 75 mm. guns.

The troops hesitated when they first burst through the gate as not a soul was to be seen and there was an ominous silence. Suddenly, the air was full of arrows and shouts of 'Allah, Allah.' So great was the shout that the leading troops halted, fearing a 'juju.' Major Marsh rushed forward to rally his men and was wounded by a poisoned arrow above the knee. He died within twenty minutes.

260 cows, 180 calves, 90 bullocks, 108 donkeys, 7 camels and 42 horses and mares were captured.

3 D.S.O.s and 2 D.C.M.s were awarded.

It would seem again from the numbers involved that more were eligible for the clasp than is shown in the reference books.

File W.O. 100/392 shows 4 officers and 45 other ranks 1 N.N.R., 11 officers and 258 other ranks 2 N.N.R. and 7 civilians and doctors, making a total of 328, were eligible for this clasp.

W.O. 100/392 under A.O. 1717 shows:—

General Lugard and other officers	34
Political Department	39
Egyptian Army Officers	8
Mounted Infantry, N.N.R.	101
Nos. 1 & 2 Bty. N.N.R.	223
No. 1 Bn. N.N.R.	702
No. 2 Bn. N.N.R.	514
S.N.R.	259
Lagos Bn. W.A.F.F.	304
	2,184

N. NIGERIA 1903-04

Army Order No. 1 of 1906 awarded the clasp 'N. Nigeria 1903-04' to all officers and under the command of Capt. G. C. Merrick, D.S.O., Royal Artillery, in the Bassa Province against the Okpotos between the 23rd December 1903 and the 12th March 1904, both dates inclusive.

This expedition was to punish the Okpotos of the Bassa Province who had murdered the Resident Capt. O'Riordan, and the police officer Mr. Amyatt-Burney, and their escort on the night of the 16th/17th December 1903.

Capt. (local Major) G. C. Merrick's force consisted of 11 Europeans, 262 other ranks and 307 carriers. He had one maxim and one 2.95" gun. He marched from Lokoja and concentrated at Ogwatcha where he organised a flying column and marched into the area where the murder had taken place.

He patrolled an area of 15 by 20 miles, fought several skirmishes in the thick bush and heavy forest and destroyed a number of villages.

Two of the N.N. Constabulary escort were rescued, some carbines recovered and generally the procedure of burning villages as a reprisal was followed.

No details are available of the enemy's casualties which must have been considerable as the British casualties consisted of:—

	Killed	Wounded
No. 1 Bty. N.N.R.	—	6
1 N.N.R.	4	10
2 N.N.R.	2	19
Carriers etc.	3	2

Merrick and three other officers received the D.S.O. while 5 D.C.M.s were awarded.

Bronze medals were issued to carriers.

File W.O. 100/392 shows the following were entitled:—

No. 1 Bty. N.N.R.	50
No. 1 Bn. N.N.R.	131
No. 2 Bn. N.N.R.	239
Northern Nigerian Constabulary	97
	517

NORTHERN NIGERIA 1904

Army Order No. 1 of 1906 awarded the clasp 'N. Nigeria 1904' to officers and men who took part in the following expeditions:—

(a) Under the command of Lieutenant S. B. B. Dyer, D.S.O., 2nd Life Guards, in the Dakka Kerri country in March 1904.

(b) Under the command of Lieutenant (local Captain) P. M. Short, Gloucestershire Regiment, against the pagan tribes who occupy the country north of Wase on the high road from the Bauchi between the 25th March and 18th April 1904, both dates inclusive.

(c) Under the command of Captain G. C. Merrick, D.S.O., Royal Artillery, against the people of Semolika in October 1904.

*(d) Under the command of Lieutenant I. G. Sewell, Royal Fusiliers,
against the Kilba tribe north of Yola in July 1904.*

These expeditions took place in widely scattered areas of Northern
Nigeria.

Dakka Kerri

The Dakka Kerri lived in the Kontagora province and had to be punished
for attacking caravans. They are described as follows in Lugard's
dispatch, 'who for the last year or so have been robbing and murdering
traders and making incursions into Kontagora country.'

Lt. Dyer's force consisted of 1 B.N.C.O., 88 other ranks, a 7-pdr.
and a maxim gun.

When they reached the town of Ebo its inhabitants refused to pay a
fine or surrender their weapons so the town was assaulted and after
considerable resistance, taken and burnt.

Wase

The Wase area was inhabited by the Yergam Pagans who made a habit
of pillaging, murdering and eating traders. A messenger sent to warn
them was killed and eaten, so the Government was obliged to send a
punitive expedition. Captain Short's force consisted of 3 Europeans,
99 ranks of the 2 N.N.R. and a maxim.

They attacked the towns of Gurkhwa and Brett. The country here
was mountainous, difficult to traverse and with very little water
available Short guarded the only available water supply and after a
few days, Skimurhe, the Headman of the Yerghum, was forced to
submit.

They then patrolled the Montoil country.

Semolika

Lt. Browne with a patrol of 50 men had been operating on the southern
borders of Kontagora to prevent the smuggling of arms and gin. On his
way back to his base he marched via the Semolika Hills, whose
inhabitants were hostile to the Government. He was attacked and
wounded, lost 6 killed and had another 17 men wounded.

Major Merrick was dispatched on a punitive expedition with a force
of 9 Europeans, 213 other ranks, two 2.95″ guns and a maxim.

The Semolikas fought hard and on the 21st April, Lieutenants
Galloway and Burnett and 9 other ranks were wounded.

Kilba

The Kilba Tribe who lived north of Yola, near the border with the
Cameroons, had closed the roads in their area and attacked caravans.

Lt. Sewell and 152 other ranks with 2 maxims was sent on a punitive patrol. Two of his men were wounded before the Kilba submitted.
File W.O. 100/392 shows the following entitled:—

No. 1 Bn. N.N.R.	214
No. 2 Bn. N.N.R.	449
	663

N. NIGERIA 1906

Army Order No. 260 of 1907 awarded the clasp 'N. Nigeria 1906' to the forces who:—
(a) Under command of the late Lieutenant F. D. Blackwood, East Surrey Regiment and Captain R. H. Goodwin, Royal Artillery, took part in the operations against the Satiru rebels near Sokoto from the 14th February to the 11th March 1906, both dates inclusive.
(b) Under command of Colonel A. W. G. Lowry Cole, D.S.O. Northern Nigerian Regiment West African Frontier Force, took part in the operations against the Emir of Hadeija from the 16th to 24th April 1906, both dates inclusive.

SATIRU

The Mimshi or Tiv, a pagan tribe, who lived in the open forest lands south of the Benue were responsible for sacking a small trading post at Abinsi. The African Hausa Manager of the store had a dispute with the locals. It came to blows and the tribesmen killed the traders, drowned their women and children in the river and then sacked and burnt the store.

Lugard's cable to the Colonial Office read, 'Mimshi tribe have destroyed Royal Niger Company's depot Abinsi. Navigation of Benue closed. Making necessary preparations for strong military expedition.' This caused consternation and Winston Churchill, the Colonial Secretary, in an uncharacteristic unjingoist mood wished to withdraw and wrote, 'I see no reason ... why these savage tribes should not be allowed to eat each other without restraint.'* In spite of this, H. Lt. Col. J. Hasler (The Buffs) led an expedition of 20 officers and 600 men, virtually all the available troops, to Abinsi, where little resistance was met.

---

* As a footnote, I must recount an anecdote told me by Malcolm MacDonald. At one time he asked a cannibal if he had ever eaten a European white man. 'No,' came the reply, 'only an American!'

At this moment trouble erupted in the far west of the territory near Sokoto. An outlaw from Dahomey named Dan Makafo persuaded Mallam Isa of Satiru, 12 miles south of Sokoto, to declare himself a Madhi. The Satirus proceeded to pay off old scores and murder their neighbours.

The acting Resident Hillary, Scott his assistant, Dr. Ellis and Lt. Blackwood, a B.N.C.O., about 70 Mounted Infantry and a maxim set out for Satiru.

The Political Officers rode ahead and attempted to harangue the rebels, but they were set upon. The Mounted Infantry rushed to their assistance but were overwhelmed by a mob of villagers armed with hoes, axes, a few spears and bows and arrows. This disastrous news was cabled to Lugard in the following form:—

'whole of C Company Mounted Infantry defeated and annihilated at Satiru ... Hillary and Scott, Residents, Blackwood, W.A.F.F. are I fear killed. Dr. Ellis severely wounded. Sergt. Slack R.A. and myself and doctor only men remaining, most urgent ... Gosling sergeant.'

'Most urgent' it was, as there was the danger that all the Mussalman Emirates in the north would revolt, and with nearly all the troops 500 miles away in the Munshi country, the position was serious indeed. However the Emir of Sokoto and others were loyal to the oaths of allegiance and Sokoto even sent a force of his men against the Satiru, but they would not attack their co-religionists.

Lugard called for reinforcements from Lagos and mustered whatever men he could find throughout the territory. Major R. H. Goodwin, R.A., with 20 Europeans and 517 African ranks, about half of whom were mounted Infantry, one 2.95" gun and two maxims advanced on Satiru, where they met the rebel force of 2,000 armed for the most part with hoes and axes, who immediately charged the British Forces.

A Sokoto Chief described the result, 'Someone gave an order, everybody fired, then a whistle blew, everyone stopped, and there was none left alive in front.'

Our forces entered the town and there did considerable execution with the bayonet. The rebel losses were put at over 1,500 killed. Mallam Isa himself was killed, but Dan Makapo and other ringleaders were captured, tried and summarily executed in the market-place, where their heads were stuck up on spears.

The Emir of Sokoto then demolished Satiru and cursed the site.

The reinforcements from the S. N. Regiment and Lagos Bn., who were halted at Zungeru, did not take part in the battle, but both the 1 and 2 N.N.R. took part in the action.

Three D.S.O.s and four D.C.M.s were awarded.

HASEITA

The Emir of Hadeija, east of Sokoto, had been in a state of sullen rebelliousness and non-cooperation.

Colonel A. N. G. Lowry Cole, Commandant of the N.N.R., led an expedition against the Emir consisting of: Artillery—2 Europeans, 33 O.R.s and two 2.95″ guns; Mounted Infantry—10 Europeans and 100 O.R.s; 1 N.N.R.—8 Europeans, 166 O.R.s and one maxim; 2 N.N.R.— 8 Europeans and 165 O.R.s. In addition, the Madeija garrison under Captain E. B. Macnaughton mustered 81 M.I. and 250 infantry. These troops had been forced to remain there during the Katiru operations.

The Emir was ordered to surrender, but he struck the messenger, defied the British and manned the walls, which like others in the north, were very formidable fortifications.

The Emir's cavalry, drugged with bhang and alcohol, charged the troops with their lances, but the N.N.R. entered the city through an unoccupied gate and five hours of street fighting in a temperature of 115° Fahrenheit then followed. Cavalrymen in chain mail with lances charged riflemen with the inevitable result, as the senselessly brave warleaders had never before met troops armed with magazine rifles.

The Emir, his sons and very many others were killed. The British casualties were negligible.

Awards included a C.B. for Lowry Cole and two D.C.M.s.

Thus ended at last the independence of the centuries' old military power of the Hausa and Fulani.

File W.O. 100/392 shows:—

Civil officers, native clerical and political staff	27
W. A. Medical Services	6
No. 2 Bty, Artillery N.N.R.	112
No. 1 Bn. N.N.R.	553
No. 2 Bn. N.N.R.	182
No. 3 Bn. N.N.R.	326
Northern Nigerian Constabulary	166
	1372

WEST AFRICA 1906

Army Order No. 102 of 1908 authorised the clasp 'West Africa 1906' to all officers and men who operated in the Chibuk country under the command of Lieut. P. Chapman, D.S.O., The Royal Fusiliers, (City of London Regiment) between the 12th November and the 4th December 1906, and all men who operated in the same district under the command of Lieutenant E. J. Wolesley the East Lancashire

Regiment, between the 22nd December 1906 and the 17th February 1907.

It will be noted that although this action spanned two years the clasp was issued for 1906 only and not 1906-07.

This action was against a pagan tribe who fought from caverns in rocky mountains, but there is no dispatch giving details.

Captain Suffolk received the D.S.O. and five D.C.M.s were issued.

OPERATIONS IN 1907 AND 1908

In operations of 1907 and 1908 in Nigeria 9 W.A.F.F. D.C.M.s were awarded. Northern Nigerian Gazette 31/3/1909 'the operations have not appeared to be of sufficient importance to warrant the publication of the despatch in the London Gazette or the grant of a medal.'

Part II

East & Central Africa

122

CHAPTER NINE

List of Medals & Clasps

Authority, Army Order or Gazette	Clasp	Reference Roll	Very Approx numbers issued
	BRITISH CENTRAL AFRICA (MALAWI) NORTHERN RHODESIA (ZAMBIA)		
66/1895	Central Africa	WO 100/76	see text
135/1899	Central Africa 1894-98	WO 100/76	997
L.G. 26466/1893	Lake Nyassa 1893	ADM 171/46	60
L.G. 26466/1893	Liwonde 1893	ADM 171/46	33
133/1902	B.C.A. 1899-1900	WO 100/91	846
90/1916	Nyasaland 1915	Unbound	907
	SOMALILAND THE MAD MULLAH, SOMALILAND (SOMAL REPUBLIC)		
133/1902	Somaliland 1901	WO 100/	420
185/1904	Somaliland 1902-04	100-105 &	1000s
	Jidballi	ADM 171/56	1000s
29/1911	Somaliland 1908-10	ADM 171/56	2350
90/1916	Shimber Berris 1914	Unbound	2400+
17/1921	Somaliland 1920	Unbound	2400+
	THE OGADEN SOMALS IN JUBALAND (SOMAL REPUBLIC)		
L.G. 26466/1893	Juba River 1893	ADM 171/46	44
125/1899	1898	WO 100/90	1386
133/1902	Jubaland	WO 100/91 &	
		ADM 171/56	1600
222/1916	East Africa 1913-14	Unbound	1141

| 95/1919 | Jubaland 1917-18 | Unbound | 925 |

THE COASTAL AREAS OF KENYA

L.G. 26657/1895	Witu 1890	ADM 171/46	1350+
L.G. 22657/1895	Witu 1893	ADM 171/46	320+
L.G. 26816/1897	Mwele 1895-6	WO 100/76 & ADM 171/46	—

UGANDA

66/1895 100/1896	Unyoro and Mruli	WO 100/76	see text
29/1899	Lubwas	WO 100/90	see text
	Uganda 1897-98	WO 100/90	see text
254/1900	Uganda 1899	WO 100/90	434
234/1902	Lango 1901	WO 100/390	490

THE NANDI & KISII

133/1902	Uganda 1900	WO 100/91	893
102/1908	East Africa 1905	WO 100/394	499
277/1906 92/1909	Nandi 1905-06	WO 100/390	2085

THE TURKHANA IN THE NORTHERN FRONTIER DISTRICT

353/1917	East Africa 1913	Unbound	265
353/1917	East Africa 1914	Unbound	264
17/1917	East Africa 1915	Unbound	491
51/1920	East Africa 1918	Unbound	1130

THE KIKUYU & EMBU

102/1908	East Africa 1902	WO 100/394	68
102/1908	East Africa 1904	WO 100/394	131
102/1908	East Africa 1906	WO 100/394	284
15/1955	Kenya	Not yet available	

Rolls are now located at Kew.

CHAPTER TEN

The Local Forces

THE KINGS AFRICAN RIFLES (K.A.R.)

Malawi (formerly British Central Africa or Nyasaland), Uganda and Kenya (formerly The East African Protectorate) each established their local armed forces independently, according to their particular needs.

THE CENTRAL AFRICAN RIFLES

Hostilities with Mkonde commenced in 1887 in the North and in the following year Captain F. D. Lugard (Norfolk Regt.) arrived and reorganised the levies into 3 companies of Atonga, Ankonde and Mambwe tribesmen.

The Arab Slave Traders, with the Yao and Angoni tribes, dominated Central Africa and the infant British Administration had to seek the help of the Indian Government and borrow sepoys from the Indian Army, who, with local levies, formed the Central African Rifles (C.A.R.)

In 1891 Capt. C. M. Maguire (2 Hyderabad Lancers) recruited 40 Sikhs from the 23rd and 32nd Pioneers and 40 Sowars from the Hyderabad Lancers, but the latter soon had to be replaced by more sepoys as their horses died of horse sickness from tsetse fly bites.

These troops wore Indian style uniforms whilst the askaris uniform was standardized as a black fez, khaki blouse and short loose khaki knicker-bockers. This, however, was special parade dress and normal duties were performed wearing a blue cotton fatigue dress which was made by the men themselves.

Troops from Zanzibar also joined the C.A.R.

By 1893 the C.A.R. comprised 3 British officers, 200 Sikhs and 150 native regulars with varying numbers of irregular tribal levies.

In 1894 further Sikhs from varying units were recruited from India.

In 1900 the force was renamed the Central African Regiment and then consisted of 6 coys of 120 rifles each and a Sikh contingent. A battery of 4 7-pdr. Mountain Guns and 2 9-pdr. field guns was also attached.

The formation of 2 C.A.R. was begun in 1899, with an authorised establishment of 19 officers, 1 warrant officer and 1084 African ranks. This battalion was almost immediately posted to Mauritius.

THE EAST AFRICAN RIFLES

The Sultans of Zanzibar employed mercenaries from the Persian or Baluch Coasts and in 1870 the Sultan's forces numbered some 2750 ranks stationed mostly in Zanzibar and at the Coastal ports of Lamu and Mombasa.

In 1877 Lt. Matthews, R.N. formed a force of about 300 Zanzibaris. He was later seconded to the Sultan's Forces as a Brigadier General and by 1880 his command numbered 1300 askaris, who were armed with snider rifles and bayonets and wore uniforms of blue jackets and white trousers with red fezzes. These men were known as 'Viroboto,' or fleas, to the Swahili of the Coast.

The I.B.E.A. Co. sought the assistance of the Sultan for maintaining order on the coast and in 1882 Matthew's troops were in action against Mbaruk of Mwele. This arrangement was not satisfactory and troops from around Delhi in India were recruited as Police, but even so, I.B.E.A. Co. had still to be dependent on Zanzibar for troops for some time.

At this stage the Coastal strip was little more than the starting point for the 'English Road,' a track which ran for 820 miles from Mombasa to Kampala in Uganda, to traverse which 90 days was allowed for an average journey. Posts were set up along the route at Machakos, Fort Smith and Eldama Ravine but the protection of caravans or convoys mostly devolved on the armed porters who accompanied these expeditions and who frequently deserted, more often than not taking their firearms with them. These firearms later came into use against the British in the many punitive expeditions undertaken in East Africa.

In 1894 a militia was raised from the Mkamba tribe who live in the Machakos area. They were uniformed in Khaki drill which by this time had become standard dress.

In 1895 the I.B.E.A. Co.'s troops totalled 866 men, of which 255 were Sudanese, and later that year, the Treasury, who were now responsible, sanctioned an establishment of 300 Punjabis, 100 Sudanese, 300 Swahilis and a mixed force of 200 men. Two years later the force still only totalled 1,120 men.

At that time the pay was Rs.18/- per month for the Indian Sepoy and Rs.16/- per month for the African Askari. In 1898 the Indian contingent was reduced in number and Swahilis recruited instead and the strength of the regiment was reduced to 1,000 rifles.

Later a Camel Corps under command of Lt. L. R. H. Pope Hennessey (O.Lt.Inf.) was formed to operate in Jubaland.

In 1899 a draft of 300 Sudanese was recruited in Egypt and by 1900 half the strength of the E.A.R. were Sudanese but by then some Kenya tribes were also being enlisted.

UGANDA RIFLES

In the mid 19th Century, Egypt had extended her Empire southwards and established an Equatorial Province to the north of Uganda. The Mahdi's rebellion in 1884 left certain Sudanese troops of the Khedive of Egypt's army isolated in Uganda, under the command of Selim Bey, an enormous and corpulent Sudanese 'giant' who had been Governor of Mruli.

Capt. Lugard, now in I.B.E.A. Co.'s service, arrived at Kampala in 1890 and found the country on the brink of civil war. The development of the Uganda Rifles is inextricably mixed with these civil wars and the mutiny of the Sudanese Troops which followed.

Lugard originally had 70 Sudanese recruited in Egypt when he left Mombasa but by the time he arrived in Uganda he had only 50 Sudanese and Somalis and 270 Porters. In 1891 he was reinforced by 70 Sudanese and 100 Swahili, a term denoting a person of mixed Arab Negro origin who was a Mussalman by religion.

Their armament included 2 maxim guns, one of which invariably jammed. Shortly afterwards reinforcements brought up his strength to 7 Europeans and 650 men.

Although Lugard laid down a scale of uniform no supplies of cloth and accoutrements were available and the askaris campaigned in whatever clothes they could get.

In spite of these shortcoming Lugard estimated the cost of a company of 120 Sudanese whose other ranks drew Rs.4/- per month at £1,000 per annum including rations, family allowances and clerical establishment, which compared with a cost of £1,300 for a company of Zanzibari troops, as these troops drew Rs.13/- per month. This pay was received in trade goods, more often than not pieces of white calico called marakani. The cost of the caravan establishment required to service 100 ranks was reckoned at about £1,250 per annum.

By 1896 the uniform adopted was a blue serge tunic and breeches with a blue jersey and khaki trousers for active service. They were armed with the Martini Henry rifle.

In 1891, however, Lugard realised that the forces at his disposal were quite insufficient to meet the tasks confronting him and he approached Selim Bey for help.

This officer, who had been separated from his Khedive for years and whose troops were unpaid, flatly refused to change his allegiance, saying he had grown grey in the service of his Khedive and nothing would induce him to change his allegiance or his flag for which he had risked his life a hundred times. Eventually Selim agreed to serve with the British but under the Egyptian flag.

Some Sudanese, 932 with arms and 1,153 without, were thus recruited and were divided into 2 Regiments.

Lugard's policy was to establish a line of small forts to separate the Buganda from the territories of the Kaberega of Bunyoro, and as these posts were manned by Selim's troops, the Egyptian flag flew over these forts.

In 1893 the British, fearing a Muslim inspired revolt from Bunyoro, withdrew some of Selim Bey's troops from the Forts. This led to a difference of opinion and the British then tried to force the Sudanese to enlist into their army, requiring them to take an oath of allegiance to the Queen. When they refused some of the troops were disarmed and disbanded and Selim Bey was convicted of treason and deported.

One of Sir G. Portal's first acts was to direct Major Owen to take over into service of Government the Soundanese. (Blue Book Africa 1893 No 4).

During Colville's administration the Soudanese were augmented by the enlistment of 500 who had marched from the Congo Free State.

In 1894 a formal protectorate was proclaimed over Buganda.

In 1895 the 'Uganda Frontier Force' was created by the 'Uganda Rifles Ordinance.' These troops were administered by the Foreign Office.

In 1897 some of the Sudanese mutinied: others were disbanded in the course of the campaigns against the Mutineers. It was therefore proposed to recruit Indian troops to form the backbone on the Uganda Rifles.

In 1900 the Regiment numbered 36 British officers, 21 British sergeants—employed as drill instructors and as commanders of the maxim gun sections—1,952 native officers, N.C.O.s and men, with an Indian contingent of 6 British officers and 402 other ranks.

The Central African Rifles, Uganda Rifles and the East African Rifles were amalgamated into the Kings African Rifles on the 1st of January 1902, as under:—

1st (Central Africa) Bn.—8 Coys formerly 1 C.A.R.

2nd (Central Africa) Bn.—6 Coys formerly 2 C.A.R. The Indian Contingent was reduced to 1 Coy in 1906 and disbanded in 1913.

3rd (East African) Bn.—7 Coys and 1 Camel Coy. All Indian Troops returned to India in 1900.

4th (Uganda) Bn.—9 Coys formerly the African Coys of the U. R. Reorganized in 1904 to include 2 Coys I.O.R.s. which were disbanded in 1913.

5th (Uganda) Bn.—4 Coys formerly the Indian Contingent of the U. R. disbanded in 1904.

6th (Somaliland) Bn.—Formed 1904 All African, in 1905 converted Indian, in 1908 Somals introduced and Indian Element reduced to 2 Coys. All disbanded 1910.

The Somaliland Indian Contingent took over from 6 K.A.R. in 1910, and operated with the 'Somaliland Camel Constabulary.' This was regularised in 1915 as the 'Somaliland Camel Corps' and the last Indian Company thereof was disbanded in 1922.

All troops were still armed with the obsolescent .450 Martini Henry Carbine but these rifles were shortly to be replaced with the Enfield .303 Rifle. 7 pdr. mountain guns, hotchkiss and maxim machine guns also formed the armament of the K.A.R.

The regulation dress was a khaki drill blouse, khaki knickers, blue jersey, blue puttees and a fez, black for he 1st and 2nd Bns., red for the rest. Askaris were not encouraged to wear boots, although sandals were allowed on operations.

The 7 K.A.R. was formed as a temporary Kaisar Wartime unit incorporating the Zanzibar African Rifles and the Mafia Armed Constabulary.

SOMALILAND CAMEL CORPS

In 1885 Britain declared a Protectorate over British Somaliland following Egypt's withdrawal in 1884. At first administered by the Government of India but passed in 1904 to the Foreign Office and a year later in 1905 to the Colonial Office.

Originally Berbera was garrisoned by a company of Indian Troops from the Aden Garrison. Later a Somali levy was formed but regular forces comprised most of the troops which operated.

In 1912 the Government authorised the formation of a 'civil' camel constabulary limited to 150 rifles and in 1914 the constabulary was reorganised on Military lines.

Lt. Col. T. Astley Cubitt was to command the renamed Somaliland Camel Corps which consisted of 3 companies mounted on ponies and camels and was regularised by the Camel Corps Ordinance in March 1915. There were 3 Camel Companies (one Indian and 2 Somali), one Pony Company and 400 Indian Army volunteers.

Following Capt. Gibb D.S.O., D.C.M.'s murder in 1922 one of the Somali Companies was disbanded and replaced by Yaos from Nyasaland and the establishment was A Coy Camel Somali, B Coy Camel Yaos, C Coy Pony Somali and 1 Coy ex 1/1 K.A.R.

In 1923 the responsibility of the S.C.C. came under the Inspector General of the K.A.R.

In 1929 mechanisation was proposed and 1 Coy was mechanised. With the advent of the war against Italy in 1940 and the invasion of Somaliland the British Forces were evacuated but the Somali were sent to their homes with their arms.

Some on the return of the British in 1941 were re-employed and in 1943 the Somaliland Scouts were formed but following a mutiny at Burao the Somaliland Camel Corps was disbanded.

CHAPTER ELEVEN

British Central Africa

HISTORICAL NOTE

The British were only concerned with the Government and development of the mainland of East and Central Africa for a comparatively short period of its history—say 75 years. Indeed until the mid 19th century the centre of the Continent, 'Darkest Africa,' really was virtually unknown.

The British Explorers Burton and Speke reached Lake Tanganyika in 1858 and from there Speke marched north to Lake Victoria and in a later expedition discovered the sources of the Nile in Uganda in 1862.

At this time slave trading was the main and flourishing Industry of Africa and it has been estimated that up to 100,000 East African natives were killed or captured each year. The entrepreneurs of this trade were Arabs and Arab half breeds with the enthusiastic co-operation of the local African chiefs.

The evangelist Livingstone reported on the extent and evils of the slave trade in Central Africa, which resulted in the establishment of a Mission Station in Nyasaland in the 1860s, but this first venture failed and it was not until 1875 that the Free Church Mission became firmly established with the foundation of the Livingstonia Mission on Lake Nyassa.

These missionaries were the Pioneers of British settlement in Nyasaland.

A. J. Swann, a contemporary writer, describes the situation thus:

'The whole of the East Africa Coast and the Interior was either in the hands of native chiefs, Arabs or marina half castes, who had all one object and whose ambition was to sell and transport to the coast as many of the inhabitants as they could possibly capture.'

'Besides those actually captured, thousands are killed or die of their wounds and famine, driven from their homes by the slaveraider. Thousands perish in internecine wars, waged for slaves with their own clansmen or neighbours, slain by the lust for gain which is stimulated by the slave purchasers. The many skeletons we have seen amongst the rocks and woods all testify to the awful sacrifice of human life which must be attributed directly or indirectly to this trade of hell.'

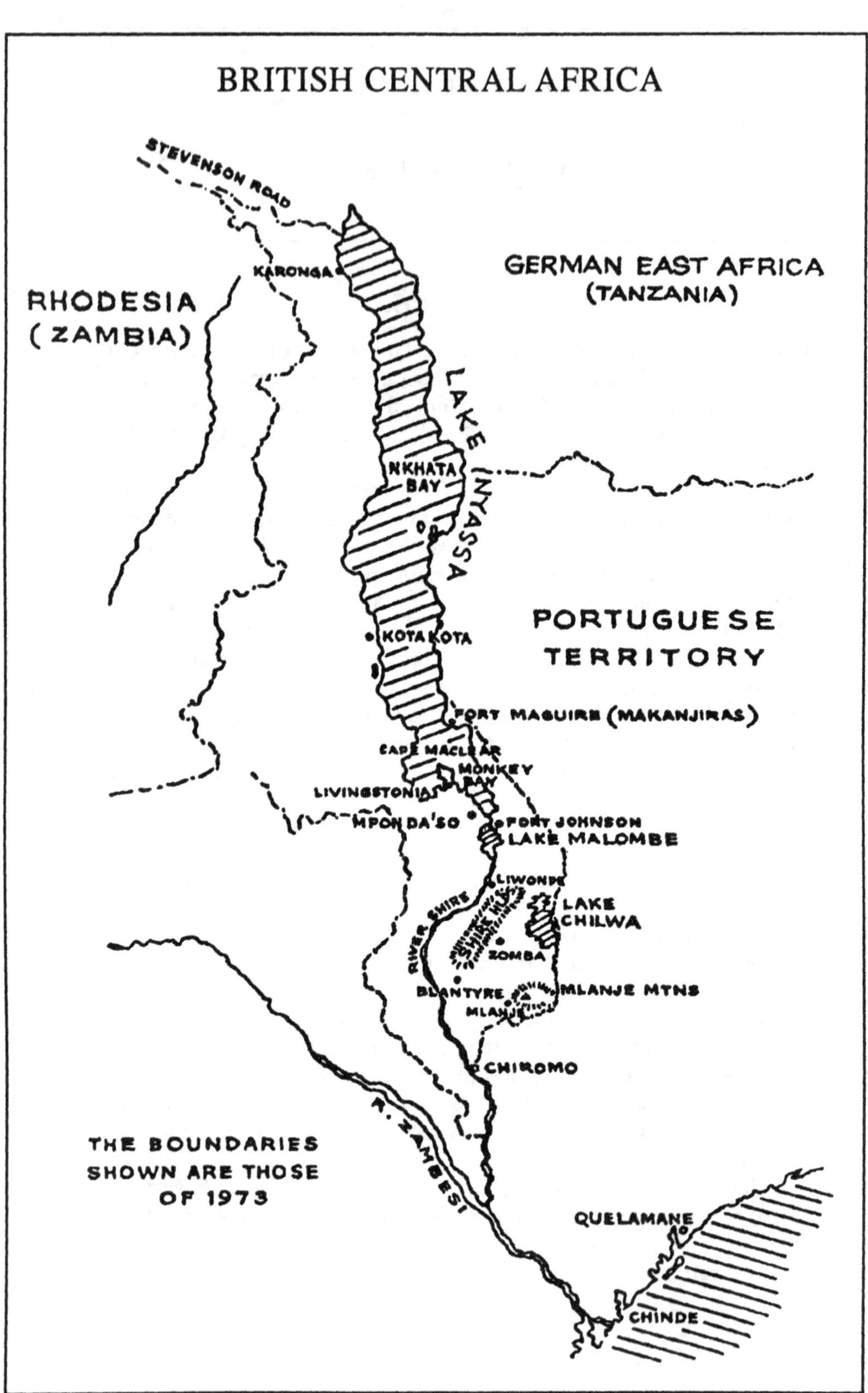

BRITISH CENTRAL AFRICA
STEVENSON ROAD
RHODESIA (ZAMBIA)
KARONGA
GERMAN EAST AFRICA (TANZANIA)
LAKE NYASSA
NKHATA BAY
KOTA KOTA
PORTUGUESE TERRITORY
FORT MAGUIRE (MAKANJIRAS)
CAPE MACLEAR
MONKEY BAY
LIVINGSTONIA
MPONDA'SO
FORT JOHNSON
LAKE MALOMBE
LIWONDE
LAKE CHILWA
RIVER SHIRE
SHIRE HILLS
ZOMBA
BLANTYRE
MLANJE MTNS
MLANJE
CHIROMO
R. ZAMBESI
THE BOUNDARIES SHOWN ARE THOSE OF 1973
QUELAMANE
CHINDE

There was also what Lugard described as 'the tyranny of the dominant tribe ...'. He describes the Angoni 'the terror and curse of all this country they seemed possessed with a lust for carnage only and kill man woman and child without distinction.'

He remarks that 'each district has its own dominant tribe' and wonders whether people in England will presently begin to realise that the Arab Slave Trader is not the only curse of Africa.

In modern times, the activities of a despot like Amin in Uganda make one realise all too well that the tyranny of the dominant tribe unhappily still exists in Africa.

In 1884 the African Lakes Company was founded.

In 1891 after some hostilities with Arab slavers, a protectorate was proclaimed over Nyasaland. In 1895 it became a Crown Colony called 'British Central Africa' and in 1907 the name was changed to Nyasaland. It is now named Malawi.

Further events are described in the North End War.

Malawi became independent in 1964.

THE SLAVERS WAR

The North End War took place before the occurrence of the expeditions for which the Central African Medal was awarded, but details are included as they are relevant to subsequent events.

Lake Nyassa, originally discovered by the Portuguese, was visited by Livingstone in 1859. In 1875, The Free Church Mission of Scotland founded the Livingstonia Mission at Cape Maclear on the lake. Trade followed, when in 1879, the African Lakes Company was formed. In 1883 a British Consul was appointed and in the following year the African Lakes Company built a trading station at Karonga, at the north-west end of the lake, which was placed under the charge of Mr. Montieth Fotheringham, the Company's agent. Pictures from his book show that the Karonga station was a small and somewhat primitive affair beside the lake shore with only a breast-high, mud-brick wall for protection.

A Church of Scotland missionary settlement was established at Blantyre, so called after Livingstone's birth-place in Scotland. The missionaries, who tended to quarrel with the Portuguese, whose colonial methods they disliked, nevertheless behaved with a certain degree of brutality themselves. One hundred lashes was a frequent punishment for fornication or robbery, and one native died after receiving two hundred and seventy lashes for stealing some beads which were later proved to have been removed by a European trader. However, this batch of missionaries was dismissed and the mission reminded that they were catering for the spiritual needs of their flock, not running a penal settlement.

The Arab slavers had divided up the land between themselves, with the Jumbe of Kota Kota allied to the Angoni tribe in the west, Mponda allied to the Yao tribe in the south and Makanjira based on the eastern shore of the lake.

J. F. Elton, who died on safari in Africa 1877, wrote in his diary:—
21.8.77 Mponda's district is entered. We found Mponda rather (as is customary with him) the worse for pombé—the native beer.
It was difficult to obtain a hearing but Mponda later said 'If slave trading is unlawful you must find someone to buy my ivory.'

Lugard describes the Arab slavers as, 'The mongrel, woolly-headed slaver of the interior, whose mother was a poor slave girl and whose sire may have been the fortieth cousin of a Muscat Arab—all alike are called Arabs.' The slavers' troops, known as 'Ruga Ruga,' were described as brigands who attached themselves to slavers and cost their masters little as their pay was a minimum share of the plunder and uncontrolled license among the captives. They often wore necklaces of human teeth and must have been very disagreeable characters.

The arrival of the British at Karonga was a threat to the slavers' livelihood and the three Arab slavers, Mlozi, Kopakopa and Msalema therefore built themselves stockades about seven miles from Karonga to consolidate their hold on the northern end of the slave route. Mlozi was a half-bred Arab; Kopakopa and Msalema, his two colleagues, were ordinary Swahilis, indistinguishable from natives, but calling themselves 'Wasungu' or 'White Men.'

In 1887 Mlozi massacred a number of Wankonde tribesmen and as a result the Karonga station had to be abandoned. Both Swann and Montieth Fotheringham described how the half-caste Arabs had fallen on the tribe at the Kambwe Lagoon, giving no quarter. Village after village was burnt and the fugitives hunted out of the hills and driven into the lagoon amongst the reeds which were set on fire. There was no escape and as the flames spread the fugitives either had to leap into the lake to be eaten by crocodiles or shot down by the Arabs waiting for them on the bank. Nowadays there are few crocodiles left and it is safe to bathe in the lake, but a hundred years ago, crocodiles were common and a real hazard.

In 1888 Montieth Fotheringham returned to Karonga with a force of eight Europeans and five hundred tribal levies, half of whom were armed with guns. He wore down the Arabs by vigorous patrols and eventually defeated them in a battle. However, there was so much plunder in the form of ivory, cloth and powder to be taken, that Fotheringham's levies just disappeared back to their villages with whatever they could carry.

Captain F. D. Lugard (Norfolk Regiment) was on leave in the area and volunteered his services to command a relief expedition. He enlisted a force of about 220 tribesmen, one third of which were armed with muzzle loaders, one third with breech-loaders and the rest spears. He was determined to attack the slavers' stockades which were formidable affairs. Their walls consisted of double rows of poles about 10–14 feet high, closely bound together, with earth rammed between the two rows of poles to make a solid, bullet-proof wall. Along the top there was often added a mass of thorn bushes to make climbing over more difficult.

An incendiary device designed by Moir, one of the European volunteers, consisted of a fire dart discharged from a 12 bore muzzle loading rifle. This missile was twelve inches long and made of solid bamboo with wings of gunpowder-soaked bark cloth. Surprisingly enough, not only could the projectile be fired without bursting the gun, but it did work and set fire to the thatch roofs of the slavers' houses.

Lugard's scratch lot of troops was really insufficient to tackle the slavers' strongholds with their ditches and walls over 14 feet high, and in a night attack on Kopakopa's stockade, Lugard was shot and wounded in both arms and chest while he was climbing over the wall leading the assault. The attack was repulsed and Lugard was lucky to survive. One European and five natives were killed and another European and nine Africans wounded.

Although the British were not powerful enough to capture the slavers' stockades, their patrols did succeed in dominating the surrounding countryside.

In January 1889, an Armstrong 7-pdr. screwgun arrived as a gift from the Nyasa Anti-Slavery and Defence Committee. To obtain permission to ship this gun duty free through Portuguese territory via the Zambesi River required the direct intervention of Lord Salisbury, the Foreign Minister in London. This was the original screwgun of Kipling's song.

It had a muzzle velocity of 1440 ft/sec. and a range of 4,000 yards, but was not particularly effective against stockades, as the high velocity projectiles passed through the walls and burst beyond. There was much trouble with the friction tubes which often misfired.

Msalema's and Kopakopa's stockades were bombarded but not reduced. But then smallpox attacked the slavers, who were already in a poor way and reduced to eating rats and they were only too ready to come to terms.

In October 1889, H. H. Johnston, the British Consul, made a treaty with the slavers allowing them to retain their stockades. The arrangement however was more of a temporary truce than a peace.

The command of the lake was prerequisite for successful campaigning and the first steamer to arrive was the 'Ilala' which was brought out by the Livingstonia Mission of the Free Church of Scotland. The 'Ilala,' named after the place where Livingstone's heart was buried, was built by Yarrow & Hedley of Mill Wall at a cost of £1,600. She was a single screw vessel, length 55 feet, beam 10 feet and displaced 21 tons. She took ten weeks to build and no single piece weighed more than 50 lbs. Her plates were bolted together not rivetted. She was assembled on the Zambesi and steamed up river to Matiti, taken to pieces and head-loaded past the cataracts of the Shire River. Eight hundred porters were employed to carry her in pieces for sixty miles, for which job they received as pay six yards of unbleached calico (marakani). They reassembled the Ilala which then steamed up the Shire River and sailed onto the lake, where the captain shut off engine and the crew all sang a hymn.

In 1891, when the Protectorate was proclaimed, H. H. Johnston was appointed Commissioner with headquarters at Zomba.

He made arrangements for the establishment of an armed police force with a nucleus of Indian troops loaned by India.

Captain C. M. MacGuire (2 Hyderabad Lancers) raised a force of seventy volunteers from the Indian Army, from the 23rd and 32nd Pioneers and the Hyderabad Lancers, and now commenced the series of small campaigns for which the Central Africa Medal was awarded. Throughout these early days both MacGuire's and Johnston's efforts were handicapped both by lack of support from H.M.G. and lack of funds to provide the wherewithal for an effective army.

Central Africa Medal Without Clasp

Army Order No. 66 of 1895—
'The Queen has been graciously pleased to approve a medal being granted to those forces employed in Eastern and Central Africa as specified below.

Mlanje (Chimkumbu) .... July–August 1891
Makanjira.. October–November 1891
Kawinga... November 1891
Zarafi.. January–February 1892
The Upper Shire................................... January–February 1893
Mlanje (Nyassera & Mkanda) .... August–October 1893
Makanjira... November–December 1893
Chirandzulu .... December 1893

The medal will be the same pattern as that granted for Ashanti but will bear no clasp and will have a distinctive ribbon.'

Brief details of these small expeditions are as follows:—

EXPEDITION AGAINST CHIMKUMBU IN THE MLANJE MOUNTAIN AREA JULY–AUGUST 1891.

This Yao slaver lived in the mountainous country around Mlanje where he dominated the peaceful Njanja tribe.

A punitive expedition was sent against him after he had attacked two European coffee farmers. After a few days hard fighting Chimkumbu fled but his brother was captured.

Captain MacGuire, Lt. B. L. Sclater, Vice-Consul Sharpe, 2 volunteers and 70 other ranks from the Hyderabad Lancers, and 23rd and 32nd Pioneers received the medal.

1ST EXPEDITION AGAINST THE SLAVER MAKANJIRA ON LAKE NYASSA OCTOBER–NOVEMBER 1891

Elton describes Makanjira as 'a slight, well made and well bred man and was simply arrayed in a checked Muscat cloth of silk mixed with cotton and bordered with a gold fringe.'

This campaign also involved Mponda, who was interfering with the passage of steamers on the Shire River.

Captain MacGuire with seventy-eight Sepoys, ten Zanzibaris and eighty Makua tribesmen and a 7-pdr. gun started to build Fort Johnston on the opposite bank of the Shire to Mponde's stronghold. From this base MacGuire moved off to attack Makanjira's stockade as a punishment for his having captured some Government messengers.

Mponda however captured and enslaved the fugitives from this action, whom he later refused to surrender. His town was then bombarded and his warriors driven out. This caused another Yao chief, Zarafi, who lived twenty miles to the east, to think discretion was the better part of valour and to sign a peace treaty with the administration.

Johnston, Maguire and the Sikhs embarked on the Domira and set off for Makanjira's settlement. Arriving early one morning they found a yelling crowd of Yaos on the Beach and immediately dispersed it by their first shell. Johnston landed with a few Sikhs under cover from Maguire's fire from the steamer, but was forced back to the ship with one or two casualties. The next morning, Maguire landed in force and after hard fighting in which several Sikhs were severely wounded, took all Makanjira's defences and destroyed his town and dhows.

No less than 8 I.O.M.s 3rd class were awarded for this action—4 to the Lancers; 3 to 23rd and 1 to 32nd Pioneers.

The Domira was a small steamer built by Matthew Paul & Co. Dumbarton, and arrived shortly after the Ilala on Lake Nyassa, having been reassembled at Matope on the Shire. She was single-screw, length 89 feet, beam 13½ feet and displaced 67 tons.

Expedition against Yao Chief Kawinga November 1891

Kawinga lived north-east of the Zomba Range and this punitive expedition was another made to release slaves. The Chief's stronghold on top of a hill was attacked by Captain MacGuire, but the attack was beaten off and MacGuire was wounded in the chest. This was enough for Kawinga who sued for peace. The Sikhs suffered several casualties.

The Vice-Consul Mr. Buchanan accompanied the expedition and 69 Indian other ranks participated. 4 I.O.M.s were awarded for the action.

After the Kawinga expedition MacGuire returned to Fort Johnston to complete its building as it was intended to garrison the post permanently.

Action at Kisangule 15–21 December 1891

Whilst he was there information was received that two of Makanjira's slaving dhows were in a creek. MacGuire embarked on the Domira. Brig. H. Bullock describes the action:

'Maguire set off in the Domira with a few Sepoys. Landing with a small force of 28 men on December 15th he was about to demolish the dhows when Makinjira, with 2,000 followers, attacked him and forced his party back to the beach. There he found his boat had been wrecked by a storm which had arisen, and the Domira herself in endeavouring to come as close as possible inshore, had stuck on a sandbank not far off the beach.

'After three Sikhs had been killed, Maguire told the others to wade out to the Domira whilst he and a few men as rearguard kept off the Yaos with the bayonet. Almost at once he was shot dead, and the European Master and Engineer of the Domira were wounded, but the remainder of the Sepoys embarked safely, whereupon they defended the stranded steamer against all comers for three days.

'Makanjira under a flag of truce then agreed, in return for a ransom— which was paid—to give up the bodies of Maguire and the three Sepoys and to assist in refloating the ship; but instead by an act of treachery, he put to death the two remaining unwounded officers, the European 1st Engineer and the Parsi Doctor.

Once again the stranded Domira was besieged, this time for 5 days till at last the Sepoys and Sowars under the direction of the wounded 2nd Officer managed to get her off the sandbank into deep water, where she used her 7-pounder with great effect on Makanjira's men dancing around their camp fires.'

The Domira managed to get away, firing into the enemy mass with her 7-pdr. The total casualties were four Europeans and nine soldiers killed, two Europeans and nine natives wounded.

This last action did not apparently qualify its participants for a medal.

Cecil Rhodes was a friend of Captain MacGuire's brother and provided £10,000 to provide the Administration with funds to defeat Makanjira and avenge his brother's death.

I.O.M.s were awarded to

> 1015 Sowar Anwar Khan, 1 Lancers
> 2557 Hav Nand Singh, 23 Pioneers
> 1909 Nk Isar Singh, 23 Pioneers
> 1179 Nk Jhanda Singh, 32 Pioneers.

EXPEDITION AGAINST ZARAFI, JANUARY–FEBRUARY 1892

Soon after MacGuire's death, Zarafi, who had entered into a treaty with the Administration a few months before, attacked Fort Johnston.

The Commissioner, H. H. Johnston, hurried north with some British volunteers and his remaining Sikhs. A siege was raised and some of Zarafi's villages on the plains were burnt, but it was not possible to bring him under control.

The Angoni offered to help Mr. J. G. King (a volunteer), who had been left in charge of the Fort, and with one hundred of their warriors, thirty Zanzibaris and thirty-five Sikhs, he set off to attack Zarafi.

On 8th February, 1892 he was defeated near Zarafi's fortress, lost six sepoys, was himself wounded and had to abandon the 7-pdr. gun. No British Military Officer participated in this Campaign. Johnston then had to remain on the defensive until the arrival in June of Captain C. E. Johnson (26 Sikhs), the New Commander, with a draft of 10 Sikhs, followed by Lt. C. A. Edwards with 100 Jat Sikhs sepoys to relieve the original contingent. Later in 1893 Lt. W. M. Manning arrived from India with 100 more Sikhs.

At this time naval reinforcements arrived, as in 1891. The Admiralty had ordered three small gunboats, H.M.S. Dove, Adventure and Pioneer from Messrs. Yarrow & Hedley of Poplar. H.M.S. Dove was a 25-ton side paddle steamer, length 65 feet, beam 14 feet and with a draft of only 8 inches. The Africans called her 'Chikapa,' which being translated from the Yao dialect means, 'The circular motion imparted to the buttocks in dancing.' Her main role was to navigate the Shire.

The H.M.S. Adventure and Pioneer were single-screw, 75 feet long, beam 12 feet and each displaced 35 tons.

These boats were tested on the Thames, taken to pieces and shipped out to Chinde at the mouth of the Zambesi. Here they were placed in lighters belonging to the German Anti-Slavery Society's expedition which was also proceeding to the lake. They were then towed by H.M.S. Herald (Lt. C. K. Robertson, S.N.O. Zambesi) to Ishikwawa, three hundred miles from the sea. From there the ships were dismantled and taken by porters to Mpimbi, eighty miles further up. Here they were reassembled, and by June 1893, all three ships were commissioned and the White Ensign flew on the lake.

The Herald and Mosquito were stern wheel paddle steamers which had been brought to the Zambesi in 1890 when there had been considerable drama with the Portuguese before they were allowed to proceed. Unlike the other small ships these were built in flotable sections which were bolted together in the water. They were a rush job and took twenty-five days to build. Before accepting them, The Admiralty ordered a test to see how long it would take to put them together. It took six hours, forty five minutes, and the job was over and the ships taken to pieces again before most of the Admiralty's inspecting personnel had arrived for the 'test.' These vessels were primarily river gunboats to be used against the Portuguese, for it appeared that they were about to annexe the Shire Highlands of British Central Africa. These was considerable diplomatic activity of the Victorian type—i.e. British fleets off Lisbon and Mozambique, but eventually the Portuguese gave way and agreement was reached without hostilities.

EAST & WEST AFRICA MEDAL

LIWONDE 1893

EXPEDITION AGAINST THE YAO CHIEF LIWONDI ON THE UPPER SHIRE, JANUARY–FEBRUARY 1893

This is the only clasp ever awarded which is named after an individual enemy, who is shown in the medal roll as 'Chief Liwondi,' and in contemporary literature as 'Robber Chief Liwondi.'

There was a serious outbreak of slave raiding on the Upper Shire River, and one of Chief Liwondi's caravans even raided near Zomba and captured some boys there.

Captain Johnson went after Liwondi and released the boys though the slavers escaped. The Commissioner Johnston then attacked and took Liwondi's town Malawi. There was general widespread unrest in the district and Johnson moved to where the trading steamer Domira had grounded on the Shire, though a local press report states that the Domira had presumably been fitted up as a gunboat for the occasion. Here he met with stubborn resistance and had to fortify himself in a small stockade, but eventually the Domira was refloated.

In the meantime Lt. G. S. Q. Carr of the H.M.S. Mosquito landed twenty-eight officers and men, and with some ten white volunteers marched 130 miles, dragging their Nordenfelt machine gun. The volunteers appear to have been under command of Baron von Eltz of the German Anti-slavery Society, and they brought a Hotchkiss gun with them.

News of what was happening had also reached Blantyre and the relief column was joined by Lt. Robertson of H.M.S. Herald and they marched together to relieve Commissioner Johnston. Liwondi was defeated and sued for peace.

File ADM 171/46 shows a total of 33 R.N. personnel on the roll and later both Carr and Robertson were awarded the C.M.G. for their services.

Robertson also received the clasp Lake Nyassa, Order of Osmanieh 4th class and M.V.O. (1905).

The High Commissioner, H. H. Johnston, C.B.; the H.M. Consul, Mr. A. Sharp, Baron von Eltz and 7 other volunteers received the Central Africa Medal.

EXPEDITION AGAINST NYASSERA AND MKANDA IN THE MLANJE AREA, AUGUST–OCTOBER 1893.

Nyassera of the Yao tribe attacked Captain Johnson in his quarters and wounded him.

Lt. Edwards led a punitive expedition and subdued Nyassera and when further trouble was given by Mkanda, another Mlanje chief, he defeated him too.

British volunteers accompanied this expedition.

2ND EXPEDITION AGAINST MAKANJIRA, NOVEMBER 1893 TO JANUARY 1894.

EAST & WEST AFRICA MEDAL

LAKE NYASSA 1893

This clasp was only granted to naval personnel. Army personnel received the Central African Medal.

Captain Johnson and two officers now had 200 Sikhs, 150 natives and a varying number of irregular levies at their disposal and they set about tackling Makanjira, who had been intriguing with a rebel called Chiwauru, who had overthrown the Jumbe of Kotakota on the East side of the lake and established his stronghold three miles from the lake shore at Kisamba.

Johnson and Edwards marched with 113 Sikhs and some Makua tribesmen and attacked Kisamba, a village surrounded by an eight foot high wall. Covering fire from a 7-pdr. gun had little effect and eventually the Sikhs charged and took the place but suffered some casualties. Chiwaura was killed and many slaves were released. The 2 Gunboats, H.M.S. Mosquito and Pioneer, were supported by the Domira and Ilala as troop carriers.

H.M.S. Pioneer and Adventure then took part in a combined operation against Makanjira's mother, Kalunda, a slave trading chieftainess, who with her warriors was shelled out of her village at Rifu Bay. Makanjira sent a dhow to rescue his mother but his was sunk by the Pioneer. Kalunda then surrendered.

Barges were borrowed from the civil authorities and fitted out a Monkey Bay. On the 19th November the troops were landed at an entrenched camp set up near Makanjira's town, later to be called Fort MacGuire.

The gunboats' Nordenfelts shelled the town and after five hours fighting the warriors gave up.

Makanjira fled to Portuguese territory, but returned later to attack Fort MacGuire where he was again defeated.

Then in January 1894, when Makanjira attacked the new fort in great force, he was defeated with heavy loss by Capt. Edwards.

No. 2188 Havildar Bulaku Singh (45 Rattrays Sikhs) was awarded the I.O.M. in this action.

The naval personnel on H.M.S. Adventure and Pioneer received the clasp 'Lake Nyassa 1893.' File ADM 171/46 shows twenty seven Royal Navy personnel on the rolls and thirty three Tindals, Seedies and Dom. 3 class. It would seem therefore that the total number of clasps awarded was approximately sixty, though 3 medals were returned to the Mint and two seamen were each issued with 2 duplicates presumably because even 80 years ago the medals were 'pricey.'

The Sikhs were eligible for the Central Africa Medal as were the High Commissioner, Consul and volunteer Mr. E. J. Glave, the explorer. Casualties were 1 Sikh and 2 irregulars killed, with nine wounded.

On 1st April 1894 H.M.S. gun vessels Dove, Adventure and Pioneer were handed over to the civil administration of British Central Africa.

Ref, ADM 1/7218 F.O. 17 Nov. 1894:—

Report by Lieut & Cdr A. T. Gurney of Pioneer at Fort Johnston.

The R.N. crews were paid off on 31/3/1895 and the ships turned over to B.C.A. Administration.

Sub.Lt. P. Cullen, R.N.R. Local Rank Comdr.

Sub.Lt. E. L. Rhodes and W. B. Williams local rank Lt., also Mr. Rendal and Mr. Gray Engineer of Yarrows.

4 POs to act as QMs from pensioners. These personnel had left UK on 31/1/1895.

It is obvious that the gunboats continued as fighting ships as these personnel won the clasp.

EXPEDITION AGAINST CHIRODZULU, DECEMBER 1893

No details of this expedition have been found.

Capt. C. E. Johnson, Lt. W. H. Manning, H.M. Consul A. Sharpe, and 63 Indian other ranks were entitled.

Roll W.O. 100/76 is confusing but indicates approximately 1200 were entitled to the Central African Medal, split between British Central Africa and Uganda.

Officers	16
Volunteers	23
Sikh Volunteers	235

The Sikh volunteers medals were engraved in India and some recipients had by then been promoted.

The Volunteers who qualified were:—

	Mlanje	Makanjira	Kawinga	Zarafi	Upper Shire (Liwondi)	Mkanda	Makanjira	Chirandzulu	
The British High Commissioner H. H. Johnston C.B.	+				+	+	+		
H.M. Consul A. Sharpe C.B.					+	+	+	+	
Vice-Consul J. Buchanan		+							
Vol. J. G. King			+						Severely wounded
Vol. A. B. Watson			+						Severely wounded
Vol. J. M. Bell					+				
Vol. T. H. Lloyd					+				
Vol. H. B. Bradshaw					+				
Vol. E. J. Glave					+	+			
Vol. E. L. Thompson					+				
Vol. F. J. Whicker					+				
Vol. A. Urquhart		+							Severely wounded
Vol. W. Keiller		+							Severely wounded
Vol. J. L. Nicoll					+				African Lake Coy.
Vol. R. Cranshay					+				
Vol. H. C. Marshall	+								
Vol. G. Hoare					+				
Vol. G. Stephenson					+				
Vol. L. M. Fotheringham					+				African Lakes Co
Vol. S. Steblecki					+				A Polish Gentleman
Vol. G. A. Taylor						+			
Baron von Eltz					+				
Vol. R. B. McAuslan					+				

The Medal Rolls shows a printed list of Sikh Personnel and Deccani Mussalmen totalling 306. General Letter No. 20 of 10th February, 1897 para 6, Encl. No. 1, Ref. 65/CA/137 India Office.

A medal to Sepoy Bela Singh 1 Regt. Inf. P. F. F. appeared in Seaby's catalogue in April 1977.

The table on p. 166 gives details of entitlement.

Central Africa Medal

CENTRAL AFRICA 1894–98

Army Order No. 135 of 1899 extended the award of the 'Eastern & Central Africa Medal' to forces employed in British Central Africa between 1894 and 1898 as specified below:

Operations for which granted	*Date of operation*
1. At and near Fort Johnston	*January, 1894*
2. Expedition against Kawinga	*March, 1895*
3. Expedition against Matipwiri	*September November, 1895*
4. Expedition against Zarafi	
5. Expedition against Mponda	
6. Expedition against Makanjira	
7. Expedition against Mlozi	*December, 1895*
8. Expedition against Mwazi	
9. Expedition against Tambala	*January, 1896*
10. Expedition against Odeti and Mkoma	*October, 1896*
11. Expedition against Chikusi	
12. Expedition against Chilwa	*August, 1897*
13. Expedition against Mpezeni	*January February, 1898*
14. Southern Angoniland Expedition	*April, 1898*

The clasp inscribed 'Central Africa 1894–1898' will also be given for these services.

OPERATIONS AROUND FORT JOHNSTON, JANUARY 1894.

No details of this expedition have been discovered. Apparently no officers participated.

2ND EXPEDITION AGAINST CHIEF KAWINGA OF THE YAOS, MARCH 1895.

The Yao Chief Kawinga decided that it was time to drive the British out of the Shire Highlands and raided near the Scottish Mission.

Corporal W. Fletcher (Royal Engineers) was sent off with a handful of men and his camp was attacked by 2,000 of Kawinga's warriors, but when some Atonga reinforcements appeared Fletcher sallied out and routed the enemy. In the meantime, Captain Manning had discovered a way to Kawinga's hilltop fortress, which he took after a two day siege.

Kawinga escaped and later changed sides and became loyal to the British.

2ND EXPEDITION AGAINST ZARAFI, AND CHIEFS MPONDA, MATAPWIRI AND MAKANJIRA, SEPTEMBER–NOVEMBER 1895.

Lieut. C. A. Edwards (35 Sikhs) with 65 Sikhs and 230 Askaris then captured Zarafi's stronghold, recovering the 7-pdr. gun lost in 1892. Zarafi fled to Portuguese territory.
Matapwiri was tackled next. In September a mixed force set out from Forts Lister and Anderson, and approaching Matapwiri's village by night achieved complete surprise and an easy victory.

Then it was Makanjira's turn to be defeated by the further expedition comprising 9 British Officers, 3 volunteers including the Governor, 7 R.N. personnel, 122 Sikhs and 43 soldiers of the B.C.A. Rifles.

This ended the conflict with the Yao Chieftains and the country round the southern end of the lake was by now peaceful and garrisoned by small detachments in the various forts and strong-points sited at strategic points throughout the area.

EXPEDITIONS AGAINST MLOZI, 'SULTAN OF NKONDE' AND HIS ASSOCIATES MSALEMA AND KOPAKOPA, DECEMBER 1895.

Mlozi, who had been quiet since the 'North End War,' began raiding the country round Karonga for slaves and threatened Karonga. A parley was held under a flag of truce, but Mlozi refused to negotiate with the British.

A force of 6 Officers, 100 Sikhs and 300 Askaris was therefore assembled at Fort Johnston and taken by steamer to Karonga. The voyage up the lake took a week's sailing.

On 1st December Lt. H. Coape-Smith (I.S.C.) posted detachments of sailors from the gunboats around Mlozi's stockade. He then bombarded Msalema's stockade which was easily captured. Kopakopa's stockade was similarly disposed of, and the forces then commenced bombarding Mlozi who replied with small arms and cannon fire. It was in the monsoon and the thatch of the huts would not catch fire. On the third day Mlozi showed a white flag, but he refused to come to terms and the attack was resumed.

Swann in his work describes the battle thus:—
'The champions of freedom were at death-grips with slavers of the most ruthless type. The fierce onslaught of our Indian troops was

met with a stubborn defence worthy of a better cause, for these people fight like demons behind entrenchments.'

Mlozi's own house was shelled and he was wounded. This incensed his followers who rushed out to attack their besiegers. Their sortie was repulsed and the Sikhs counter-attacked, swarming over the walls to capture the town and Mlozi.

The Sikhs lost one killed, whilst the Askaris lost three killed and six wounded, in comparison with the besieged's losses of over two hundred killed.

A photograph of Mlozi appears in Moir's book 'After Livingstone.' He is a benign looking old Arab gentleman with a long white beard, sitting cross-legged on a carpet and wearing a white cap and kanzu. He looks exactly like the Arab traders of today who still sit outside their 'dukas' in old Mombasa or Malindi, but the only difference is that the background in Mlozi's case is a line of eight Roga Roga armed with guns in front of a stockade. This picture reminds me of a pre-independence conversation with a similar Arab in Malindi referring to the imminent departure of the British. He said, 'Once you go we will control them again'—fortunately a false surmise.

The next day Mlozi was tried, condemned and hanged for the murder of the Wankonde.

The Arab's stockade was then destroyed and the execution brought an end to the Slavers War.

EXPEDITION AGAINST MWAZI, DECEMBER 1895.

The expedition against Mlozi was soon followed by an expedition against Mwazi, an ally of the Arabs. Mr. A. J. Swann, the Political Officer at Kotakota, with 2,400 auxiliaries and Lt. E. Alston (Coldstream Guards) with 40 Sikhs and 149 native troops, with their own arms and ammunition, set out to attack Mwazi. On arrival at his village and after some preliminary abuse such as 'We'll send you back to the Lake to eat fish' etc., Lt. Alston fired three volleys, had the charge sounded on the bugle and attacked. In the ensuing battle nine soldiers were killed and five wounded, some from gunshot and others from arrow wounds.

Mwazi was defeated but escaped, but Saidi Mwazungu, the ringleader of the murders of 1891, was captured. After a long trial he was executed for the murder of Dr. Boyce at Kotakota.

298 rifles, gunpowder and £1,200 worth of ivory was captured.

EXPEDITION AGAINST PARAMOUNT CHIEF TAMBOLA OF THE ANGONI, JANUARY 1896.

The Angoni were the descendants of a Zulu clan which had crossed the Zambesi River in 1835, from where they had moved northwards, raiding and looting and adding to their wealth of cattle and slaves until they settled in the area west of Lake Nyassa.

The Angoni were armed with the standard Zulu weapons—an oval shield, throwing assegais, a heavy stabbing spear and a knobkeri. Whilst they did not live in fortified villages like the Arab slavers, their villages were sited in inaccessible positions difficult to surprise.

In January Edwards led an expedition against Tambola, the paramount Angoni chief, who lived on the top of a table mountain, three of whose sides were precipices, whilst the fourth and only approach route was covered by large boulders. The village was commanded by artillery from another hilltop and was taken after a bombardment but Tambola escaped and joined the Chewa raider Odete.

EXPEDITION AGAINST THE CHEWA CHIEF ODETE WHO WAS HARBOURING TAMBOLA AT MOUNT CHIRENJE AND MKOMA, OCTOBER 1896.

After a five day march from Kotakota, Manning, with a force of 24 Sikhs and 80 Africans, used a little-used track up the mountains, surprised Odete and stormed his stronghold. Odete then submitted.

EXPEDITION AGAINST CHIKUSI, OCTOBER 1896.

Manning then joined Captain Stewart at Zomba on a punitive expedition against Chief Chikusi of the Angoni. Stewart's forces comprised four officers, 58 Sikhs, 198 African troops and a 7-pdr. gun. They attacked Chikusi on 21st October. He was captured, tried, found guilty, condemned and shot.

EXPEDITION AGAINST CHIEF SERUMBA OF THE ANGURU TRIBE TO THE SOUTH OF LAKE CHILWA, AUGUST 1897.

This chief's tribe had been guilty of highway robberies. The punitive expedition under Manning, of four Officers, 51 Sikhs and four companies of Africans, met little opposition and Serumba was captured, his village burnt and a fine levied.

EXPEDITION AGAINST CHIEF MPESENI OF THE CHIMPINGO IN THE LUANGWA VALLEY IN NORTH EASTERN RHODESIA (ZAMBIA) JANUARY–FEBRUARY 1898.

The North Charterland Exploration Company had tried to expropriate a large tract of Angoni tribal land by a fraudulent concession obtained by the German Karl Wiese. The old Chief Mpeseni could no longer control his son Singu and the younger warriors, who were reputed to number between 10,000 and 25,000 and these now threatened the Company which appealed for help.

Captain H. E. J. Brake (Royal Artillery) wit a force of seven officers, 118 Sikhs and all six companies of the C.A.R. with artillery and maxim guns was sent to relieve Fort Jameson.

He was then attacked by 600 warriors who after one volley, in which they lost twenty dead, would not stand up to the British force. This attack was followed by another by more tribesmen which was similarly routed.

The expedition then rounded up cattle, burnt villages and generally chased the Angoni until Mpezeni surrendered and his son Singu was captured. After a trial Singu was shot before the assembled headmen.

Colonel Callwell's Stationary Office publication 'Small Wars' points out that:

'During the operations against the Angoni Zulus in the Mpeseni country in 1898, an insignificant force of 400 local troops with two guns, by dint of rapid marches and resolute initiative, completely broke up a numerous and warlike tribe, seizing their cattle when they fled and attacking them whenever they attempted to stand.'

EXPEDITION TO SOUTHERN ANGONILAND, APRIL 1898

Captain F. B. Pearce (West Yorkshire Regiment) and Lt. Brogden took a strong patrol to Dedza where a force of armed Angoni were reported. Order was restored without much fighting.

W.O. 100/76 shows the following entitled to the clasp:

Sir H. H. Johnston and Civilian Volunteers			21
British Officers (Army 23, Navy 4)			27
British N.C.O.s			2
Naval Personnel, Boatswains, Engineers, Sarangs and ABs			16
B.C.A. Rifles	1st Bn.	2nd Bn.	
Indian Havildars	—	2	2
Askaris	629	90	719

Of the 2 Bn., 37 were Sergeants, 28 Corporals, 20 L/Cpls and only 5 Privates.

Sikh Contingent of the B. C. A. Rifles	213
Total	1000

Racewise the Number of Medals issued with Clasp Central Africa 1894–1898 was:

Europeans			67		
Deduct Clasps			4	=	63
Indian Rank			220		
Deduct	Clasps	19			
	Bronze	9	28	=	192
African Ranks					719

DeductClasps		3			
Returned Mint		10		13	= 706
Total	Silver				961
	Bronze				9
					970

From the above it does not seem justified to describe this clasp as rare and the commonest clasp 'Mpense' was hardly a skirmish.

Africa General Service Medal

Army Order No. 133 of 1902 authorized the African General Service Medal with clasp 'British Central Africa 1899–1900' for Imperial and Colonial Forces who were employed in the military operations in British Central Africa who took part in:

1. *The actual operations against Nkwamba, in August to October, 1899, under Captain F. B. Pearce, West Yorkshire Regiment, Tombowe to be considered the base for the operation.*
2. *The actual operations in North-East Rhodesia, against Kazembe, in September to November, 1899, starting from Karonga.*
3. *The actual operations in Central Angoniland against Kalulu in December 1900, under the command of Captain R. Bright East Kent Regiment.*

BRITISH CENTRAL AFRICA 1899–1900

Operations against the Yao Chiefs Nkwamba and Mataka in the Namwerra Hills August, October 1899.

These chiefs, who lived on the Easern Frontier of British Central Africa, had been raiding and causing trouble in both British Central Africa and Mozambique which culminated in Nkwamba ambushing two European traders and killing some members of their safari.

Captain F. B. Pearce (West Yorkshire Regiment) left Zomba with 9 Officers, 119 Sikhs, 269 Askaris of 1 C.A.R. and after very little resistance destroyed Nkwamba's village. However Mataka moved into Portuguese territory where he caused them trouble for some time.

Operations to aid Chief Kasembe of the Bemba Tribe near Lake Mweru, North East Rhodesia, September–November 1899.

This chief had previously refused to enter into a treaty with the British authorities but now sought our aid against the last remaining slavers in his area.

Captain E. C. Margesson (South Wales Borderers) and Lt. P. C. B. Barclay (I.S.C.) with 12 Sikhs, 60 Askaris and a 7-pdr. gun marched from Karonga to Abercorn and from thence south-west to Mazebes. The slavers fled into Belgian territory and Kasembe then entered into a treaty.

The troops involved marched 1,000 miles in two months.

OPERATIONS AGAINST THE ANGONI, NOVEMBER–DECEMBER 1900.

Captain L. F. de V. Stokes (Royal Lancashire Regiment), who commanded Fort Mlangeni, led an expedition which destroyed 300 villages and killed 20 Angoni.

In spite of these reprisals a mail runner was murdered and a strong force was therefore sent out to avenge his death. More villages and stockades were destroyed and Tambola, who had escaped in 1896, was captured.

The very last slave dhow operating on the Lake was said to have been captured in 1908.

We can find no mention of Captain R. Bright mentioned in A.O. 133/1902 except in the Medal Roll.

Medal Roll No. W.O. 100/91 shows that approximately 850 all ranks were entitled to the medal and clasp.

Officers	16	
Volunteers	10	(including Mr. Alfred Sharp and a nursing Sister L. Harrison)
Staff & Medical	14	
Indian Contingent in B.C.A.		
Indian Other Ranks	187	
Askaris	619	

Both these detachments belonged to the 1st Battalion Central African Rifles, later renamed 1st Kings African Rifles, although the sepoys medals may have been named showing their parent units.

The medal rolls show a bewildering number of units that were apparently present, and the Indian list shows (current names shown)

2 Derajat Mountain Battery	2	2 Derajat Mtn Bty
2 Peshawar Mountain Battery	2	3 Peshawar S. P. Fd. Bty
12th Bombay Infantry	6	5 Bn The Grenadiers
14th Bombay Infantry	3	Mahrata Training Centre
20th Bombay Infantry	2	2 Bn Rajputana Rifles
23rd Bombay Rifles	1	4 Bn Rajputana Rifles
28th Bombay Pioneers	3	Disbanded 1933
12th Burma Infantry	8	3 Bn Punjab Regt

29th Burma Infantry	10	The Baluch Regt
30th Burma Infantry	4	The Baluch Regt
31st Burma Infantry	8	The Baluch Regt
32nd Burma Infantry	5	The Baluch Regt
1st Punjab Infantry	4	7 Bn Cokes The F. F. Regt.
19th Punjab Infantry	1	1/14 Punjab Regt
20th Punjab Infantry	2	2/14 (DCO) Punjab Regt
23rd Punjab Infantry	1	Disbanded 1933
24th Punjab Infantry	3	4/14 Punjab Regt
26th Punjab Infantry	4	2/15 Punjab Regt
27th Punjab Infantry	4	3/15 Punjab Regt
28th Punjab Infantry	4	4/15 Punjab Regt
31st Punjab Infantry	4	2/16 Punjab Regt
32 Punjab Pioneers	4	Disbanded 1933
1st Sikh Infantry	4	3 Bn F. F. Regt
3rd Sikh Infantry	5	5 Bn F. F. Regt
35th Sikh Infantry	7	Regt Centre Sikh Regt
36th Sikh Infantry	11	4 Bn The Sikh Regt
45th Sikh Infantry	11	3 Bn The Sikh Regt
14 Ferozepore Infantry	7	1 Bn Sikh Regt
15 Ludiana Sikhs	10	2 Bn Sikh Regt
3 Q O. Corps of Guides	5	9 Bn Punjab Regt

A veritable roll call of the Indian Army, but of course these units were not present as formations, but details formed part of the 1 K.A.R. (Central African Rifles) Indian Contingent. The charm of collecting medals to Indian soldiers is that their units live on in India, Pakistan and H.B.M.'s Gurkha Bde.

Unfortunately the Roll does not show in which campaign the recipients fought.

NYASALAND 1915

Army Order No. 90 of 1916 authorised the African General Service Medal with Clasp 'Nyasaland 1915' to officers, soldiers and other engaged in the operations against the rebels in the Shire Highland of Nyasaland between the 24th January 1914 and the 17th February 1915, both dates inclusive.

The Watch Tower Sect was an offshoot of Jehovah's Witnesses which originated in the U.S.A. at the beginning of the century. When adapted for African consumption it became racist and rejected the Christian way of life, emphasising ritual and adopting pagan ethics.

John Nkologo Chilembwe, its leader, was a Yao tribesman, born near the Chirandzulu Boma. He was first trained by Joseph Booth the founder of 'The Zambezi Industrial Mission' who baptised him in 1893. In 1897 they went to the U.S.A. to raise funds for the African Christian Union which was designed to prevent Africans relying on Europeans. Chilembwe parted from Booth and studied at the Virginian Theological Seminary under the auspices of Pastor W. Brown of the High Street Baptist Church. In 1900 he returned to B.C.A. via London as an ordained Missionary and launched 'The Providence Industrial Mission' (P.I.M.) also known as the 1st African Baptist Mission.

He expanded this into a substantial church and by 1911 had 6 or so Mission schools under his Control.

According to Sir Philip Mitchell, then a District Officer but later Governor of Kenya, 'his teaching at least had some appearance of Christian worship but its driving force was hatred of Europeans.' He however encouraged his followers to wear European clothes.

He extended his activities and according to current African historians preached Africa for the Black Africans and Freedom, which with the 1914–18 War brought him into conflict with the Administration. In November 1914 Chilembwe wrote to the Nyasaland Times regarding 'The voice of African natives in the present War.' Receiving no response he next addressed the Governor who reaffirmed his interests in African welfare and arranged for officials to call at the P.I.M. on 25th January but Chilembwe heard from one of his spies that he was to be arrested. So he decided to launch his uprising on Saturday 23rd when the conspirators reckoned the Europeans would all be drunk.

The plan was for a three pronged attack

1. At Magomero where the European males were to be murdered and at Bruce Estates Mr. W. J. Livingstone's head was to be cut off and brought back to the P.I.M. Mission as a sign the Revolt had begun.
2. An attack on the Mandala store at Blantyre to obtain firearms to be removed and distributed at P.I.M.
3. An uprising at Zomba.

The attack on the Livingstone's house went according to plan. The Freedom Fighters were led in by Mrs. Livingstone's trusted houseboy Linjesi Khomelera, Livingstone was murdered and decapitated in his wife's presence. Later another planter MacCormick was speared to death and the European women were removed as captives.

At the Mandala Arsenal not all the insurgents turned up. The watchman raised the alarm, only 10 guns were stolen, and the Europeans rushed over from the Club and drove off the raiders.

At Zomba those loyal to the Government prevented Simon Kadwewere from starting an uprising.

Thereafter the rebels received little support. Mrs. Livingstone was quickly rescued, there was a skirmish at Malindi where the Government forces suffered casualties, Chilembwe became a fugitive. On 3 February

1915 he was captured near Chinorampeni Hill. Modern historians say ... 'He had a Bible in his hand ... he just knelt against a tree praying ... they fired at him needlessly and then seized him ... he prayed and they shot him dead.'

The Police maintained he was shot trying to escape.

Captain L. E. L. Truscott (K.A.R. Reserve) with forth European Volunteers took 100 recruits from the Zomba depot to the Shire Highlands and scattered the rebels. Reinforcements of the 1 K.A.R. came from Karonga under command of Captain H. G. Collins and arrived just in time to finish off the remnants who were hunted down in a few days. Some were tried summarily and shot.

The medal rolls, which are still unbound, show that the following were entitled to the medal and or clasp:—

The Depot Company 1 K.A.R.—5 officers and 221 other rank
'F' Coy 1 K.A.R.—6 officers and 110 other ranks
'H' Coy 1 K.A.R.—2 officers and 80 other ranks,
Nyasaland Volunteer Reserve—226. These are shown as on patrol
    duty, on garrison duty and field duty.
Central Prison warders—6
African Transcontinental Telegraph Co.—3
Native Police, Blantyre, Chirandzulu, Fort Johnstone, Mlanje, Ncheu
    and Zomba—162
Civil list of native volunteers—30. These recipients had all rendered
    special services and the roll states remarks such as:
        'Was largely instrumental in giving the first alarm of the
            outbreak and in doing so received three spear wounds ...
            succeeded in conducting to safety his employer.'
        'In charge of the telephone in the Concentration Camp.'
European Volunteers — 55. These included H.E. The Governor Sir
    George Smith, 3 civilian doctors, 4 nurses and a dresser and a
    number of civilians whose services are cited, such as:
        'Arrested over 201 rebels who were fully armed and raiding in
            the Upper Shire District and gave services in the demolition
            of undesirable schools.'
        'Had a special train made up and under steam for three days,
            had 400 natives brought in.'
        'Captured Peters, one of the ringleaders.'
        'Was engaged in the fighting around Chirandzulu and
            Machembo.'
        'Acted an interpreter to the Judge.'
        'Member of Firing Party.'

A total of 907 clasps appear to have been awarded and it is difficult to believe that all of these recipients saw any action.

THE CENTRAL AFRICA MEDAL: A QUESTION OF VALUES

The Central Africa Medal with Ring suspender (authorised by Army Orders No. 66/1895 and 100/1896), and with Clasp 'Central Africa 1894–98' (authority Army Order No. 135/1899), was awarded for 22 separate small campaigns against Slavers and Tribal Chiefs; Three of the expeditions took place in Uganda and the rest in British Central Africa (Malawi) and Northern Rhodesia (Zambia). Royal Naval personnel were involved in two of the B.C.A. battles, and the officers, ratings and seedies from four H.M. ships Herald, Mosquito, Adventure and Pioneer were awarded the East and West Africa Medal with Clasps 'Liwondi 1893' and 'Lake Nyassa 1893' (authority London Gazette No. 26657 of 1895). This is one of the few examples of entirely different medals being awarded for the same campaign.

The Medal Rolls for the Military, Naval and Civilian Personnel (which do not appear to be complete) are contained in File W.O. 100/76, and the Lake Nyassa and Liwondi campaigns for Naval personnel are in File ADM 171/46.

Gordon, presumably writing in the 1940s, states 'the medal with bar is very rare and that with swivel suspender is far from common.' I rather doubt that this opinion is correct, as, whilst the Medal Rolls show that approximately the same numbers of each type of medal were issued, I believe that the named medals issued of the first ring type are almost three times as rare as the clasp medal of the second type. Be that as it may, the Market believes Gordon, as at the Lovell Sale at Sotheby's, a ring suspension medal sold for £100, whilst a clasp medal fetched £210.

The relative interest between these particular two medals is extreme; the former issued to Sowar Ahmed Shah Khan of the Hyderabad Lancers has a history whilst the latter to Corporal Uladi of 1 K.A.R. is commonplace. The 22 Cavalry Sowars from the 1st, 2nd and 3rd Hyderabad Lancers and the Mazbi Sikhs of the 23rd and 32nd Pioneers were the first Regular Army troops ever to operate in East and Central Africa. The Lancers arrived in Africa with their horses, which quickly died from the effects of tsetse fly bites, but dismounted they took part in four separate campaigns against Chimkumbu at Mlanje, against Makinjira and Kawinga in 1891 and against Zarafi in 1892. They also served in the S.S. Domira as 'horse marines' at the disastrous action at Kisangule on Lake Nyassa when their commanding officer, Captain Maguire (2 Hyderabad Lancers) was killed, 2 other Marine officers wounded and 3 Indian soldiers killed. This action, presumably as it was a defeat, did not qualify the participants for the medal.

The Lancers won 5 Indian Orders of Merit (3rd class) apart from losing 3 killed, and in all during this period the Indian Troops, including the Mazbi Sikhs, were awarded 9 I.O.M.s and lost 9 killed. So Sowar Ahmed Shah Khan's medal was won by hard fighting and is also scarce by any standard.

What of the Clasp Medal sold for £210? This was issued to Corporal Uladi of the B.C.A. Rifles (1 K.A.R.). This unit earned 629 Central Africa Medals with clasp, and Corporal Uladi himself took part in only one campaign against Mpenzi in Northern Rhodesia in January/February 1898. No less than 124 of the Indian contingent and 574 askaris from the 1 and 2 B.C.A. Rifles are on the Roll for this campaign which is the commonest of the series. It appears that the only fighting that took place against Mpenzi was the repulse by one volley each of 2 separate rushes by the Angoni and none of the Military were casualties.

To revert to relative rarity, as the Rolls are incomplete it is difficult to make a precise estimate of the numbers of 1st type medals issued, particularly as the notes in the Files are confusing. It is convenient to divide the issue between recipients in Uganda and B.C.A.

It appears that probably between 700 to 900 medals were issued for the three Uganda campaigns: in Unyoro (December 1893 to February 1894), in Mruli (April to June 1894) and in Unyoro (April to May 1895). It will be noted the swivel medal was awarded in 1894 and 1895 when the clasp medal was awarded in B.C.A.

The Rolls only give a total of:—

British Officers	13
Civilians	4
Buganda Chiefs	21
Uganda Rifles	447
2nd List of Uganda Rifles (?)	111
	596
Returned	77
	673

However, a note on page 207 of the Roll states 'Rolls for the remainder have been promised by the Foreign Office,' but there is no trace of these. 77 medals were returned to the Mint so the aggregate increases to 673, still well short of the total of 900 indicated in a note on page 206 of Roll W.O. 100/76; so it seems a medal list was lost, perhaps in the Mutinies, for it must be remembered that the Uganda Railway was not through to Lake Victoria until 1901 and communications were not always first class.

There is evidence that some of the medals were presented, because Lieut. Vandeleur in his book Campaigning on the Upper Nile & Niger mentions that Berkeley, the Commissioner, held a parade of Sudanese Troops in October 1895 in Kampala for this purpose; but the Medal Roll was not apparently sent back to England by Berkeley (later Sir Ernest) until 5 February 1896, so it seems very probable that the Uganda Medals were issued unnamed, and the Rolls prepared after the soldiers had received their awards. Some of the Sudanese on the Rolls can be identified as Mutineers and also it seems likely that their medals were lost or destroyed in all the confusion and fighting of 1897–1899.

There is a record that a duplicate medal was issued by the Chief Secretary of Uganda to a Sudanese, Nafar Gadein Rehan, in 1919, and it is tempting to guess that this was named up before it was despatched from the U.K.

Unnamed medals appear on the market at fairly regular intervals and it seems probable that a hundred or so unnamed 1st ring medals may still exist. It would be extremely interesting, therefore, if some collector could provide any details in regard to the naming, if any, on medals issued for the campaign in Unyoro 1893/4, Mruli 1894 and Unyoro 1896.

The original Indian Troops loaned to the B.C.A. administration were 22 Sowars from the Hyderabad Lancers and 50 non-Jat Sikhs from the 23rd and 32nd Pioneers and these were replaced in 1893 by a contingent of 200 Jat Sikhs, the pick of the Sikh Regiments.

The names on the medals issued to the Indian Contingents in B.C.A. were engraved in India (W.O. 100/76, page 166). It is possible to say precisely, for 270 of those entitled, in which campaigns they participated, but some 35 are also shown as entitled on a printed list without specifying the campaign in the Rolls. The medal to 4134 Pte. Bela Singh, 1 Regt. Sikh Infantry, which was in Seaby's April 1977 List, is one of these. The numbers of named medals issued for Central Africa are therefore about 300.

In all in this period 13 I.O.M.s were awarded to the Indian Troops for gallantry.

The 23 civilian volunteers entitled to the ring 1st type medal included the Commissioner, the Consul, the Vice Consul and Baron von Eltz of the German Anti-Slavery Society. Two of these volunteers died and 6 were wounded so they too saw some fighting.

The attached Table A shows approximately how many participated in each campaign.

I have found no record of whether the Native Contingent's medals were named but Hayward's List of May 1976 contained a medal inscribed '2457 C.A.F.F.' Could this have been issued to an African Askari?

As stated earlier, there should be more clasped medals remaining in existence than named ring medals, as approximately 1,000 of the former were issued and also some 26 recipients of the ring medal converted them to clasps, so these should be deducted from the 300 named 1st type medals.

The Navy had some 200 personnel present who won the Central Africa Medal with clasp but I could, however, only trace one Officer in the contemporary Naval List (Lt. Comd. Gurney, H.M.S. Pioneer) so it is difficult to suggest what the rest were doing in B.C.A. at this time as H.M.S. Dove, Adventure and Pioneer were handed over to the Civil Administration in 1894.

A summary of personnel entitled to the Clasp Medal is given in List B. It will be noted that apparently only 9 Sikhs were participators in the action in January 1894 'At and near Fort Johnston' and they were without officers.

What happened to all the B.C.A.'s clasp medals, many of which must be in pair with the later A.G.S.s? Were they melted down? Obviously the market cannot be all that wrong but if the medals did come on the market as frequently as the Rolls indicate it should be reflected in the price.

Table A—Expeditions which qualified forces for the Central Africa Medal (Army Order 66 of 1895)

	MLANJE Jul–Aug 1891	MAKANJIRA Oct–Nov 1891	KAWINGA Nov 1891	ZARAFI Jan–Feb 1892	UPPER SHIRE* Jan–Feb 1893	MLANJE Aug–Oct 1893	MAKANJIRA† Nov–Dec 1893	CHIRODZULU Dec 1893	TOTAL MEDALS	List Genl. Letter 10.2.1897 Ref. 68/ Central Africa/137.
Officers	2	1	1		1	3	2	2	5	
Civilian Volunteers	3	3	1	3	10	6	3	1	23	
Native Contingent Armed Forces B.C.A.					1	2	9		12	
1st Regt. Lancers Hyderabad Cont.	7	7	7	7					7	7
2nd Regt. Lancers Hyderabad Cont.	10	10	10	9					10	10
3rd Regt. Lancers Hyderabad Cont.	5	5	5	4					5	5
23rd (Punjab) Regt. Bengal Inf. Pioneers	25	26	25	24					26	26
32nd (Punjab) Regt. Bengal Inf. Pioneers	23	23	22	20					23	23
1st Regt. of Sikh Infantry P. F. F.						8	11	4	15	21
15th Regt. of Bengal Inf. (Ludhiana Sikhs)					8	11	1	14	20	
22nd (Punjab) Regt. Bengal Infantry						4	4	2	6	8
24th (Punjab) Regt. Bengal Infantry						10	11	8	14	13
26th (Punjab) Regt. Bengal Infantry						4	4	2	5	4
28th (Punjab) Regt. Bengal Infantry						18	11	3	19	29
30th (Punjab) Regt. Bengal Infantry						13	7	4	13	16
31st (Punjab) Regt. Bengal Infantry						12	4	7	14	20
35th (Sikh) Regt. Bengal Infantry						1	15	8	15	13
36th (Sikh) Regt. Bengal Infantry						23	23	10	26	30
45th (Rattrays) Sikh Regt. Bengal Inf.						16	13	12	17	20
Indian Medical Service						2	1	1	2	3
	75	75	71	67	12	130	127	65	270	305

East and West Africa Medal with Clasps Liwondi 1893 and Lake Nyasa 1893

H.M.S. Mosquito and Herald	33	33
H.M.S. Adventure and Pioneer:		
R. N. Personnel	27	
Seedies	32	59

* Liwondi 1893   † Lake Nyasa 1893

Table B—Expeditions which qualified the forces for Central Africa Medal with Clasp
(Army Order 135/1899)

	At & near FORT JOHNSTON Jan 1894	KAWINGA Mar 1895	MATIPWIRI Sep–Nov 1895	ZARAFI Sep–Nov 1895	MPONDA Sep–Nov 1895	MAKANJIRA Sep–Nov 1895	MLOZI Dec 1895	MWAZI Dec 1895	TAMBALA Jan 1896	ODETI & MKO Oct 1896	CHIKUSI Oct 1896	CHILWA Aug 1897	MPEZENI Jan–Feb 1898	SOUTHERN ANLAND Apr 1898	TOTAL	CLASP ONLY
British Officers		2	5	8	5	9	5	1	2	1	5	3	7	6	24	2
British N.C.O.s		1	1	1	1	1	1	1	1		1			1	2	
R.N. Officers		1		2	2		2						1		4	
R.N. Personnel incldg. 7 Sarangs				13	13	7	9				1		2		16	
Volunteers		5	4	4	3	3	3	2		1	2	2	4	4	21	2
(Europeans)		9	10	28	24	20	20	4	3	2	9	5	14	11	67	4
2 (Derajat) M.B., P.F.F.			4	4	3	3	4	3	2	3	4	2	4	1	5	
4 (Hazara) M.B., P.F.F.			4	4	4	4	4	1	1		2	2	3		5	
1 Regt. Pun. Inf. P.F.F.			5	3	4	4	4	1	2		5	1	4	2	8	
2 (Hill) Sikh Inf. P.F.F.			2	6	5	4	4	1	2	2	7	3	6	1	10	
4 Sikh Inf. P.F.F.	1		4	6	7	7	4	1	1	1	4	3	6	2	10	
11 (PWO) B. Lancers			1	1	1	1	1								1	

14 (Ferozepore Sikhs) B.I.	1														1	
15 (Ludhiana Sikhs) B.I.				8	3	5	6	1	1		3	3	9	1	11	
10 (P) Regt. B.I.				2	2	3	3			2	3		1		3	
21 (P) Regt. B.I.			1			1								1	1	
22 (P) Regt. B.I.			1											1	1	
24 (P) Regt. B.I.				1	2	8	4	2	2	2	4	3	6	3	11	
25 (P) Regt. B.I.			4	1	1	1	1	4		3	1	3	7	1	9	
26 (P) Regt. B.I.				1	4	7	7	5	3	1	4	2	6	2	10	
27 (P) Regt. B.I.				1	1	5	7	4	7	3	2	2	7	4	10	
28 (P) Regt. B.I.				2	5	5	7	4	6	1	3	6	9	14	11	
29 (P) Regt. B.I.			3	4	5	7	5		2		3	2	4	1	10	
30 (P) Regt. B.I.			4	4	4	3	3	3	2	1	4		3		6	
31 (P) Regt. B.I.			2	3	2	8	9	3	5	2	3	4	8	2	12	
35 (Sikh) Regt. B.I.	7	2	13	20	13	26	13	3	4	2	10	10	18	11	50	15
36 (Sikh) Regt. B.I.			8	9	7	6	6			2	5	2	5	1	10	
45 (Rattrays) Sikh Regt. B.I.		15	16	11	13	13	4	7	1	10	7	14	6	19		
Indian Medical Service			1	1		1	1		1	1	1	1	2	3		
Agra Police													1	1		
2 BCA Rifles: Havildars													2	2	2	2
(Indians)	9	4	70	97	84	122	106	40	51	24	78	59	124	42	220	19
1 BCA Rifles		8	29	55	38	41	28	14	25	11	80	368	502	110	629	3
2 BCA Rifles		2		6	4	5	1	1	4	1	9	65	73	19	90	
(Africans)		10	29	61	42	46	29	15	29	12	89	433	575	129	719	3
Total of Troops entitled	9	23	109	186	150	188	155	59	83	38	176	497	713	182	1006	26

'Race'	Total	Deduct Clasps	Issued	Forfeit	Deduct Bronze	Net
Europeans	67	4	63			63
Indian	220	19	201		8	193
Africans	719	3	716	10		706

The Emin Relief Expedition Star

The Emin Relief Expedition 1887–89 Star was awarded by the Royal Geographical Society to the followers of the expedition led by Henry Morton Stanley to rescue Emin Pasha who was supposed to be stranded in Central Africa.

The expedition combined quite unnecessary tragedy and farce: tragedy because it was pointless for Stanley to rescue Emin anyway and rather than proceed by the same well tried slave route from the East Coast of Africa he had used in 1874 he elected to explore via the West Coast, with resultant delays and heavy casualties, presumably to obtain publicity for himself; and farce as apart from giving the notorious Moslem slave trader Tippu Tib, who controlled an area of the Congo, a bull terrier as a gift, Emin did not want to be rescued and finally when he had been persuaded to accompany Stanley to the East Coast fell out of the window at Bagamoya Club in German East Africa and nearly killed himself.

'Mehmed Emin Pasha (1840–92) was a German Eduard Schnitzer who was born in Opole, Silesia on March 28, 1840. He graduated in medicine at Berlin and was then for nine years in Turkish government service, becoming a first-rate administrator as well as a brilliant linguist. He adopted a Turkish name and in 1875 went to Khartoum to join General Charles Gordon as medical officer and in 1878 he became governor of Equatoria.

'Emin had been cut off by the Mahdi revolt of 1882 in the neighbourhood of Lake Albert, and Sir William McKinnon, chairman of the British East Africa company, launched a rescue fund and invited Stanley to lead an expedition. Finance for the expedition had been generous with £20,000 being made available, including £10,000 from the Egyptian government. Stanley offered his services as leader without fee or reward, giving up numerous lucrative engagements to undertake this hazardous mission. Stanley left England in January 1887, stopping to recruit men in Zanzibar, and when the expedition left it consisted of 9 European officers, 61 Sudanese, 13 Somalis, 3 interpreters, 620 Zambizaris as well as Tippu Tib, and 407 of his people. The expedition left aboard the 'Madura' and reached the navigable head of the Congo in June.

There Stanley left a rear column of four white men and 257 natives under Major E. M. Barttelot and J. S. Jameson, with orders to await Tippu Tib's 500 porters and to follow on should they not arrive. On June 28 Stanley led out of the camp the advance guard of Lt. W. G. Stairs, Capt. R. H. Nelson, Surgeon T. H. Parke and A. J. Monteney Jephson and 389 men. After a gruelling march through dense, unknown forest, fights with pygmies, starvation and daily deaths, they finally reached Lake Albert, 470 miles and five months later. Such were the privations that by this time they were only 174 men strong. Contact with Emin was not made until April 1888 when Stanley found to his obvious astonishment that the Pasha was very reluctant to follow him to the coast. Leaving Emin to assemble his troops, Stanley returned to locate his missing rear column. He reached Balanya, just short of Yambuya on August 17 where he discovered the remains of the column, depleted and very demoralised. Tippu Tib had not sent the porters he had promised, Barttelot had been murdered and Jameson was soon to die of fever. Returning to the lake in January 1889 Stanley learnt that Emin and Jephson had been detained by disaffected troops but eventually the party was reunited and on April 10, 1889, about 1,500 people set out, arriving at Bagamayo on the coast on December 4.'

The losses were extremely high as will be appreciated from the following figures relating to the non-European expedition members:—

	Somalis	Sudanese	Zanzibaris
Died/Killed	12	38	257
Executed	–	2	2
Deserted	–	1	191
Survived the journey	1	19	225
Started Out	13	60	675

The five silver pointed star was manufactured by Carrington and Co. hallmarked Birmingham 1889/90.

Obverse:	A circular band inscribed 'Emin Relief Expedition 1887–9' encircling the intertwined initials 'RGS'.
Reverse:	Plain, generally with a Birmingham (P) hallmark
Suspension:	Two rings (one of which is attached to the top point of the star)
Naming:	The stars were issued unnamed except for 9 issued to the Chiefs
Number issued:	200 stars were purchased (some records suggest 193) and approximately 175 were sent to the intended recipients.

Of the stars, 3 are in the R.G.S. museum. Very few of the 200 can be expected to have survived.

Poor Emin Pasha came to an untimely end back in Africa. He was approached by his assassins, who enquired if he could read Arabic. On receiving an affirmative reply they handed him their warrant instructing them to cut his throat, which they duly did.

Stanley, the illegitimate son of Welsh parents, brought up in the workhouse and later in the States, obviously had the most monumental chip on his shoulder and seems to have been generally at odds with 'officers and gentlemen' even though he was honoured by gifts from H.M. The Queen, knighted in 1899 on being appointed G.C.B. and showered with gold medals by learned societies. He was a Victorian Hero but he cannot have been a very nice person to work with.

Major Barttelot's diaries are highly critical of Stanley's ability and integrity and his brother's published letter of accusation to Stanley, included in his book, which is full of charges, including 'suppressio veri' and 'suggestio falsi' remained unanswered by Stanley.

In Stanley's newspaper articles published in instalments on the 1868 Magdala and 1873 Coomassie expeditions he is a carping and patronising critic of the British Military and what is surprising is that they still tolerated him.

The Africans called Stanley 'Bula Matari' or Stonebreaker. The African is generally perspicacious with his nicknames and Stanley's indicates that they saw through him for what he was, a ruthless man, but he took it as a compliment as it was inscribed on his tombstone.

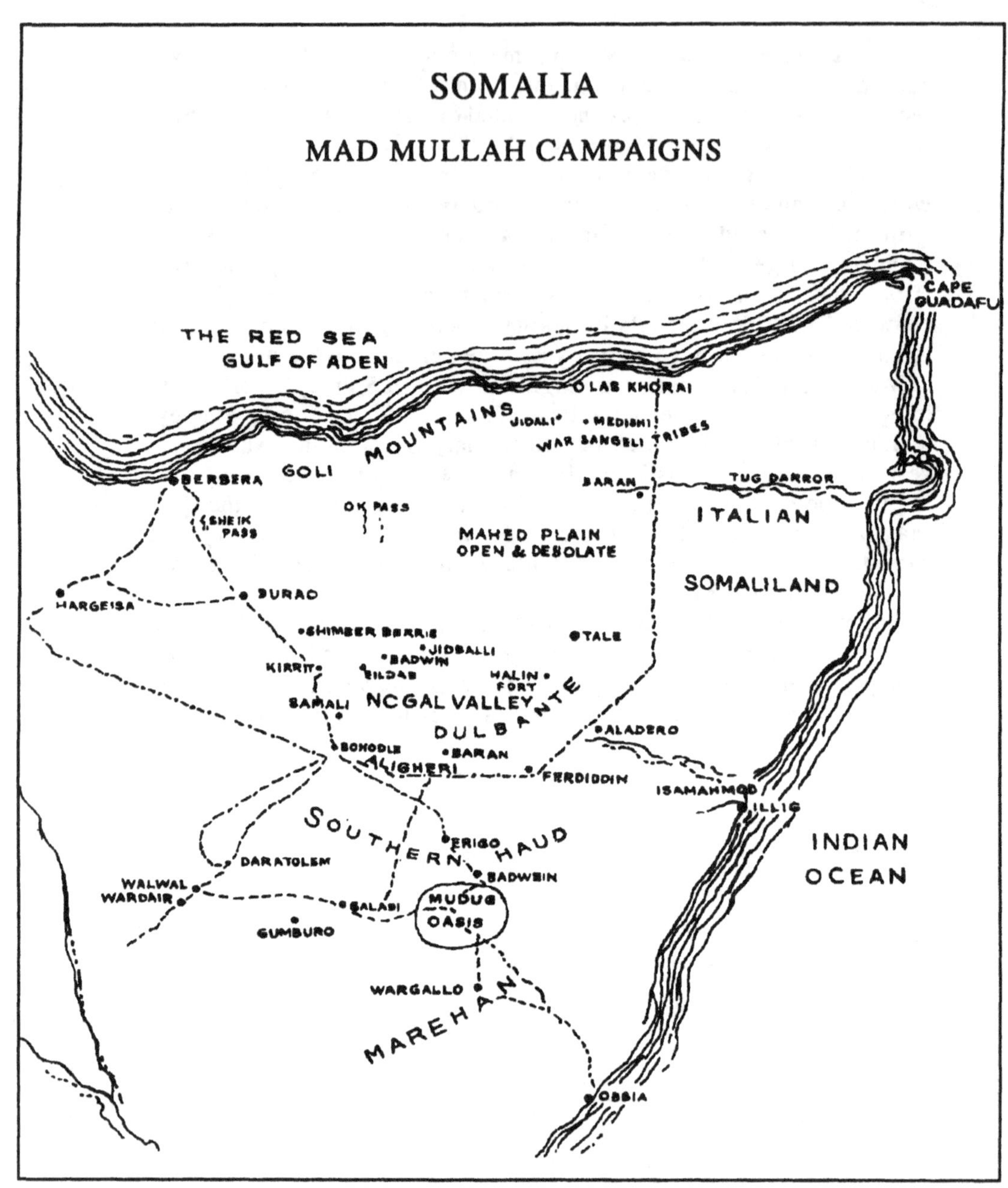

SOMALIA
MAD MULLAH CAMPAIGNS
THE RED SEA
GULF OF ADEN
CAPE GUADAFU
MOUNTAINS
LAS KHORAI
JIDALI
MEDISHI
WAR SANGELI TRIBES
GOLI
BERBERA
BARAN
TUG DARROR
ITALIAN
OK PASS
SHEIK PASS
MAHED PLAIN
OPEN & DESOLATE
SOMALILAND
HARGEISA
BURAO
SHIMBER BERRIS
TALE
JIDBALLI
BADWIN
KIRRIT
EILDAB
HALIN FORT
SAMALI
NCGAL VALLEY
DULBANTE
ALADERO
BOHODLE
BARAN
ALIGHERI
FERDIDDIN
ISAMAHMOD
ILLIG
SOUTHERN HAUD
INDIAN OCEAN
ERIGO
DARATOLEM
BADWEIN
WALWAL
WARDAIR
GALADI
MUDUG OASIS
GUMBURO
WARGALLO
MAREHAN
OBBIA

CHAPTER THIRTEEN

Somaliland

The Red Sea littoral and Somalia formed part of the Ottoman Empire which came under Egyptian administration in 1866, and was known as 'The East Coast of Africa' from Suez to Cape Guadalfui. In 1873 the Egyptian Forces occupied the port of Berbera. Berbera was a great entrepot for trade with the Harrar District of Eastern Ethiopia and the principal port on the Somali coast, whose harbour was formed by a spit of land 1½ miles in length, and was the only one where vessels could anchor in all weathers.

The British and Egyptian navies then cooperated together for the suppression of the Slave Trade in the Red Sea until the Egyptians abandoned Somalia in 1884 after the Mahdi had rebelled against them in the Soudan.

The whole area was then divided into spheres of influence by the British, French, Italians and Ethiopians. The British occupied Berbera when four of the Somali tribes placed themselves under their administration for protection against the Ethiopians. The old British Protectorate is now the Northern Province, which forms only a small part of Somalia.

The Protectorate of Somaliland was administered from India until October 1898 when control was taken over by the Foreign Office. In 1905 it came under the Colonial Office.

It is a dry and arid country with a range of hills called the Golis Mountains rising up to 5,000–6,000 feet in height and running roughly parallel and at a short distance along the northern coast. To the south of the Golis Range a large plateau stretches right across the country with a gradual fall to sea level in the south-east on the coast of the India Ocean.

This plateau is covered with belts of dense thorn jungle interspersed with open spaces where after rainfall the grazing is good.

In the dry weather much of the eastern area is uninhabitable. The areas known as 'The Haud' are a vast belt of thorn wilderness and of pasturages which are dry and virtually uninhabitable throughout the Jibal—the hot season. It has a breadth of 150 to 200 miles and stretches from near Jig Jagga to the maritime ranges of what was Italian

Somaliland. The greater part of it is in Ethiopia, but part lies along the southern part of the old Protectorate. The Haud acted as a barrier to large scale movements of flocks except when grazing was available after rainfall.

There are occasional showers of rain in April and May and again in August and September. Generally however rainfall is scanty and when the rain does come, it falls heavily, converting the dry, sandy river beds into torrents and ensures grazing in the area of precipitation for some time to come.

Water, or the lack of it, governs all tactics and movements in the country, where marches in the dry weather must of necessity be between water holes or wells.

Campaigning in the intense heat and dust through thick bush, where visibility could be down to a few yards, with insufficient water supplies and against an elusive and cunning enemy, was arduous work for both man and beast, and although there were not many battles, the medals awarded were very well deserved.

The Somalis comprise various tribes who trace their descent from certain Arab chiefs who introduced the Moslem religion to their country in the sixteenth century. These Arabs intermarried with the indigenous Galla tribes who probably moved from Arabia some time B.C. Their descendants, the Somals, are a wiry people capable of great endurance, excitable, inordinately vain, with avarice as their chief characteristic and possessing an overriding desire to loot camels. For all that 1909 description, 'Snaffles,' the artist, described the Somali as 'a cheery, good-looking lad—but a bit hard on his women,' and went on to say that the Somal is the only black man in Africa who appreciates his animals and has the uncanny Somali trick of spiriting camels away from their lawful owners.

He make an excellent though somewhat excitable soldier. They are a handsome race and the girls are beautiful—or so I am told by a former officer of the Somali Camel Corps. Their wealth was composed entirely of sheep, goats and camels (burden, milk and food).

Formerly the Somal fought as a spearman and his usual arms were a small round leather shield about one foot in diameter, a light spear for throwing and a heavier one for thrusting. Some also carried short straight, double-edged swords called 'bilawas.'

The Mijjarten tribes used clubs, while the low caste Midgans a knife and bow and arrows, but the latter were not generally found amongst the Dervish Forces.

In 1900 the tribes had few rifles, but as time went on, more and more became armed with firearms which were smuggled into the country.

As a fighting man, the Somal was not a Ghazi like the Sudanese, and although the Mullah did imbue his followers with a fanatical spirit, they never staked all in one mad rush, but inevitably attacked in several lines of extended groups with their riflemen distributed along the front. If their centre was checked the flanks wheeled inwards like the horns of a Zulu impi.

The Mad Mullah, or to give him his full title, Haji Muhammad-bin-Abdullah Hassan of the Kabi Sulamian Clan of Ogaden Somalis, was born in 1864 near Kirrit in the Dolbahanta country. By the age of 19 he had won himself the title of Sheikh for his learning and piety and acquired sanctity by pilgrimages to Mecca (The Haj). He joined the Salihaya order, became a militant reformist of this puritanical sect, and it soon became apparent to the authorities that he intended to organise a religious war against the British, for in 1899 he raided into the Protectorate as far as Burao (3,000′).

In that year he proclaimed himself 'The Mahdi' and wrote to the Consul-General at Berbera saying, 'Choose for yourselves, if you want war, we accept it.'

It was believed that the Mullah was demented, and alleged that in his younger days he had had a bone removed from the top of his head and this somewhat amateur surgery had made him insane. Hence the reason for his being known as the 'Mad Mullah.' However, his terrible behaviour involving pillage, rape and murder was quite sufficient to earn him that sobriquet whatever his actual mental condition may have been.

One curious fact is that during the twenty years he fought the British, at no time was he ever actually seen by a European.

The following clasps were awarded to the African General Service Medal for the several campaigns against him:—

Somaliland 1901	22nd May–30th July	1st Expedition Lt. Col. E. J. E. Swayne
Somaliland 1902–04	18th Jan 1902 –11th May, 1904	2nd Expedition Lt. Col. E. J. E. Swayne
Jidballi	10th Jan 1904	3rd Expedition Brig. Gen. W. H. Manning
		4th Expedition Major Gen. Sir C. C. Egerton
Somaliland 1908–10	19th Aug 1904–31st Jan 1910	
Shimber Berris 1914–15	19th–25th Oct 1914 and 2nd–9th Feb, 1915	
Somaliland 1920	21st Jan–12th Feb 1920	

In the intervening periods between the medal-winning expeditions there were numerous clashes between the Mullah's followers and the British Forces.

SOMALILAND 1901

(22ND MAY–30TH JULY, 1901)

Army Order No. 133 of 1902 authorised for Imperial and Colonial Forces who were employed in the military operations in Somaliland the medal with clasp 'Somaliland 1901' to all officers and men who took part in the operations in Somaliland against the Mullah Muhammed-bin-Abdullah under the command of Lieut.-Colonel E. J. E. Swayne, Indian Staff Corps., and to those detailed to accompany the Abyssinian force which cooperated between 22nd May and 30th July, 1901, both dates inclusive, the base for the expedition being considered as Burao.

1ST EXPEDITION

In 1900 the Mullah's men had raided the Dolbahanta tribes who live in the Nogal Valley, which is a vast open area collecting a great part of the drainage of the higher plateau around Burao, situated at an altitude of about 1,500'. This forced the Government to take action and a Local Defence Force under Lt. Col. E. J. E. Swayne (Indian Staff Corps) was raised. It was then considered that Somali levies would be more suitable than regular troops for dealing with such an elusive enemy as the Mullah since the levies would be able to live in the country.

The levies consisted of a Camel Corps of 100 men and mounted infantry under command of Major W. G. L. Beynon, and two corps of 500 infantry each under command of Capt. G. E. Phillips (Royal Engineers) and Capt. M. MacNeill (Argyll and Sutherland Highlanders).

The operational plan was that, with the cooperation of the Ethiopians in the east, the Mullah should be driven into the Dolbahanta country where he and the Aligheri tribes who were helping him would be attacked and punished.

Swayne advanced from Burao to Samala where he built a zariba which he garrisoned with a force of 470 under command of Capt. MacNeill.

He then set off in pursuit of the Mullah, who side-stepped his main body and on the 2nd and 3rd June attacked the Samala zariba. But his forces were driven off with the loss of 600 men killed and wounded compared with British casualties of 10 killed and 9 wounded.

Swayne then intercepted the Mullah's forces at Odergoeh and their retreat become a rout as they fled over the border to the Mudug oasis.

In a short time he was driven out of Italian territory by the hostility of the Sultan of Obbia and had to return to the Nogal Valley. Swayne received intelligence of this and decided to attack.

After a night march he attacked the Mullah's forces at Ferdiddin at dawn, defeated them completely and burnt down the Mullah's village. The Mullah lost over 60 men killed, whilst the British Forces lost 10 killed and 17 wounded.

Swayne then returned to Burao.

During this campaign the Mullah's losses totalled 1,200 killed and 80 captured.

Apart from fighting the battles at Samala and Ferdiddin, the levies covered 1,170 miles in three months and carried, besides their rifles, bayonets and equipment, 100 rounds of ammunition, two days' rations of dried meat and dates and a sheepskin containing a gallon of water. When required, they put up a spirited fight, which was all the more remarkable as most of the officers could only communicate with their troops through an interpreter.

The Somali Campaigns Medal Rolls 1901–1904 are contained in five volumes W.O. 100/101–105.

Those shown as entitled to the 1901 clasp comprise:—

Somali levies (6 K.A.R.)	328
Somali levies & F.F.	92
	420

More medals should have been issued but perhaps by the time the rolls were prepared, the whereabouts of many of the levies may have been unknown.

SOMALILAND 1902–04

(18TH JANUARY, 1902–11TH MAY, 1904)

These operations can be divided into three separate expeditions.

Army Order 185 of 1904 authorised The African General Service Medal to naval and military forces who served in the Somaliland operations between 18th January, 1902 and 11th May, 1904. The medal was also granted to those officers who accompanied Colonel A. N. Rochfort, C.B., C.M.G., with the Abyssinian forces.

The medal will also be granted in silver or bronze, as the case may be, to all authorised Government followers who accompanied the troops so engaged.

The clasp attached to the medal will be inscribed 'Somaliland 1902–04.'

A clasp inscribed 'Jidballi' will also be granted to those who were present in that engagement, and to those who formed part of the guard left behind during the engagement in charge of the baggage under the command of Major W. B. Mullins, 27th Punjabis.

In December 1901 the Mullah was back in British territory proclaiming a Jeddah (Holy War) to drive the British out of Somaliland and punish those who had helped them. He made several raids around Burao. At this time the Mullah's forces consisted of about 12,000 mounted men and 3,000 foot.

An ever increasing number of Le Gras rifles of 1874 pattern that had been smuggled in from the coastal ports were reaching the Mullah. As a rifles was worth five or six camels, gun-running was a lucrative trade and the British and Italian navies cooperated to try to stop the dhows smuggling these arms from Jibouti and Arabia, but without much effect.

2ND EXPEDITION

In May Swayne set off with a force of 1,200 infantry and 70 mounted troops reinforced by 300 men from the 2 K.A.R. and 60 Sikhs from Central Africa. He established a base on the border at Bohotle where he built a stockaded fort and then attacked the Mullah's fort at Halin where a number of rifles and large numbers of stock were captured.

In October Swayne advanced southwards from Baran with a force of 1,250 fighting troops and 4,000 camels into Italian territory.

On 6th October 1902 the Mullah attacked Swayne in thick dense bush at Erigo. The British formed square and were assaulted from all sides by riflemen and spearmen. There was considerable confusion when the Dervishes, as the Mullah's followers were now known, almost broke into the square when some of the Somali levies panicked and the camel transport stampeded. Some equipment, including a maxim which fell off a camel, was lost during this stampede and two officers, including Major Phillips, the 2nd in command, and 99 other ranks were killed, with roughly the same amount of wounded. In this battle Lt. Col. A. S. Cobbe (32 Pioneers) won the Victoria Cross for rescuing a wounded man under fire.

Only 66 dead Somalis were counted, although it was subsequently learnt that the Mullah considered the battle a reverse. Swayne retired to Bohotle where there was now a strong masonry fort.

The campaign involved 1,500 miles of marching and counter-marching and H.M. Force's inflicted up to 1,400 casualties on the

Dervishes and captured 2,500 camels, 1,500 cattle, 200 horses and 250,000 sheep.

3RD EXPEDITION

Colonel Swayne was recalled and Brig. General W. H. Manning placed in command of the forces which were based at Bohotle. The Somali levies, now formed into the 6 K.A.R., were reinforced by 1 Bombay Grenadiers, 150 Punjab Mounted Infantry and by the end of 1902, the British forces totalled 2,674 all ranks.

Consent was given by the Italians to operate in their territory and the plan was to land a force on the east coast at Obbia, march northwards and drive the Mullah out of the Mudug oasis area, while 5,000 Abyssinians under Fitaurari Gabri were to act as a stop to the west and a column from Berbera was to move southwards towards Mudug to corner the Mullah.

The following general instructions were issued for the guidance of the General Officers Commanding:—
1. The primary object to be the expulsion of the Mullah from Mudug.
2. The Italian Government made it a condition of their assent to the landing of a force at Obbia that operations should aim at preventing the Mullah from retiring south-west.
3. If the Mullah should retire or be driven from Mudug, he should be pursued, but not for a greater distance southwards or westwards than four or five days' march (This restriction was subsequently removed, and, after the occupation of Mudug, General Manning was practically given a free hand.)
4. An Italian officer to accompany the Obbia force as Political Officer, and his advice to be taken with regard to the future control of the Mudug district.
5. The occupation of Mudug to be reported to the Secretary of State with a view to the issue of further instructions.
6. An Abyssinian force was to occupy the eastern frontier of their own territory and act as a stop to the Mullah, if he retired in that direction.
7. The base at Obbia was to be closed by the end of April, and the Obbia column to return via Bohotle.
8. Attention to be paid to the improvement of roads and water supply in our own Protectorate.

The Somali Field Force to land at Obbia consisted of 86 British officers and 2,256 other ranks, including one section of the Lahore Mountain Battery, 8 S & M, 2 Sikhs P.F.F., Bikanir Camel Corps British and Punjab Mounted Infantry and 1, 3 and 5 K.A.R.

The Berbera column comprised 78 British officers and 2,297 other ranks—1st Bombay Grenadiers, 7 Bombay Pioneers, Indian Contingent

B.C.A. (Sikhs), 2 and 6 K.A.R. and a telegraph battalion of the Royal Engineers.

The landing at Obbia was difficult as there was no harbour and everything had to be landed through the surf. Horses and mules had to be swum ashore whilst the camels were either towed or landed in Madras surf-boats. There were insufficient baggage animals available and the Sultan of Obbia, who refused to sell the expedition any camels, was eventually deported by the Italians, whereupon camels became available for sale from the area.

In February, Manning left Obbia with a flying column for Wargallo, and from thence made contact with the troops from Bohotle. The Mullah was supposed to be somewhere to the west between the British and Abyssinian forces and for the next few weeks the troops manœuvred to the west but only captured a few thousand stock and killed some Somals.

The Abyssinians were attacked at Burhilli, near the Webi Shebeli River to the west, where there was a severe hand-to-hand fight in which the Abyssinians preferred using their swords to their rifles. They evidently enjoyed themselves, as their losses were 21 killed and 10 wounded, whilst the Mullah's men had 300 killed and 3 wounded. This unusual ratio was perhaps because it was an Abyssinian custom for each man to present himself to his commander after a fight and perform a war dance, and as he was not eligible to do this unless he had killed an enemy, perhaps he had to have his victim's head as a trophy and proof of his prowess.

The Mullah was then supposed to have gone further west to Walwal and Wardair, whence the column, under command of Lt. Col. A. S. Cobbe, V.C., proceeded to march. As Cobbe was unable to maintain his forces in the Haud indefinitely he was anxious to bring the Mullah into action.

On 17th April Cobbe was camped in a Zariba near Gumburu Hill. The Mullah's main forces were known to be in the area and Capt. H. E. Olivey (Suffolk R.) and a company of 2 K.A.R. were sent out on a reconnaissance. Olivey reported contact with the enemy and asked for reinforcements and returned to within a mile of the Zariba.

Lt. Col. A. W. V. Plunkett was sent out with orders to support Olivey and bring him back to camp. Instead, however, of returning as ordered, Plunkett moved out about another six miles from the Zariba where he was surrounded and attacked by all the Mullah's forces, consisting of some 4,000 horsemen and 10,000 foot.

The square survived until their ammunition started to run out, when the order was given to break out with the bayonet and make it back to the Zariba. Plunkett was killed and his force virtually destroyed, as out of 224, all the 9 officers were killed and only 38 other ranks survived, of which only 6 were unwounded.

Thereafter Cobbe retired to Galadi.

In the meantime, the Bohotle column of 213 all ranks under command of Major J. E. Gough (Rifle Brigade) advanced in a south-westerly direction and a base was established at Danot.

On the 23rd April, on a flat plain covered with thornbush up to 20′ high, contact was made with the enemy at Daratoleh, just north of Walwal.

Gough formed square and was attacked by some 300 riflemen and 500 spearmen for three hours. After firing 200 of the 300 rounds carried per man, the force withdrew towards Lanot.

Victoria Crosses were awarded to Major Gough, Capt. Rolland and Capt. Walker for this fight.

British casualties were 6 officers and 38 other ranks killed or wounded, whilst the Dervishes lost about 150 killed or badly wounded. Eleven rifles were captured.

By this time the Obbia line of communication had been closed and Brig. General Manning concentrated his forces at Bohotle. The May rains then made it possible for the Mullah to escape with all his forces and stock to the Nogal Valley, and thus ended the second round of this campaign, for which the honours must go to the Mullah.

4TH EXPEDITION

After these reverses, British prestige was at stake, and forces were increased to divisional strength of 6,389 all ranks and placed under command of Major General Sir E. E. Egerton, K.C.B., D.S.O. The reinforcements included 2, 4 and 6 K.A.R., Indian Contingent B.C.A. Camel Corp Battery, 1 Bn. Hampshire and the 27 Punjabis.

The order of battle was:

1st Corps—Mounted Troops: 3 coys British M.I. of 60th Rifles, 2 coys Somali M.I.

2nd Corps—1 coy Poona M.I., 1 coy Ambala M.I., Bikanir Camel Corps, 1st Brigade—1, 2, 3, 4 & 5 K.A.R., Indian contingent B.C.A. Camel Battery.

2nd Brigade—1 Hampshire Regiment (wing), 27 Punjabis, 57 Sikhs, 28 Mtn. Battery.

Lines of Communication—101 Grenadiers, 107 Pioneers, 6 K.A.R., 2 coys 3 S & M.

An advance base was prepared at Kirrit, a convenient spot from which to strike either at the Nogal Valley or the Bohotle area and by the end of October one brigade was concentrated at Bohotle and the other at Eil-Dab. Brig. Gen. Manning was ordered to occupy Galadi, which he did with a garrison of 300 or so all ranks. The Navy meantime

175

demonstrated off Obbia with a view to making the Mullah, who was at Aladereo, believe another landing was contemplated.

In January 1904 a Field Force was assembled at Badwein where a large bivouac protected by barbed wire was established.

A reconnaissance party of mounted infantry discovered the enemy in force at Jidbali and were involved in a skirmish, when Lt. Carter was awarded the Victoria Cross for rescuing a dismounted trooper of the Poona M.I.

JIDBALLI

(10TH JANUARY, 1904)

On the 10th January, 1904 General Egerton marched out over open country with a force of 2,200 regulars and 1,000 irregulars to Jidbali, a distance of about 12 miles. This force included 2 coys British M.I., 2 coys Indian M.I., 1 coy Somali M.I., Bikanir Camel Corps, Tribal Horse, Gadabursi Horse, Sec. 28 Mountain Battery, Hampshires, 27 Punjabis, 52 Sikhs, 1 coy S & M, 2 maxim gun detachments, 107 Pioneers and K.A.R. (500 men).

General Egerton advanced to the attack at Jidbali and formed the Infantry in double echelon from the centre with mounted troops on the flanks.

Unexpectedly, the Dervishes, who were not led by the Mullah in person, stood to fight and opened a heavy but inaccurate fire on the British troops, who formed square. The Dervishes then attempted to charge the square, but the disciplined fire of the troops, the shells of the Mountain Battery and the fire of the maxims halted them, and after a few minutes, they broke and fled, pursued by the Cavalry.

The Mullah's forces were estimated as being in excess of 6,900, mostly of the Dolbahanta tribe, and suffered over 700 killed, whilst the British casualties only totalled 27 killed and 37 wounded.

Thereafter followed mopping-up operations which had the effect of scattering the Mullah's followers, who now suffered many desertions.

The accepted spelling of this place is Jidbali and this was in current usage in the Somaliland Armed Forces, even though the War Office Map of 1922 spells it 'Jidballeh.'

It was thought that the Mullah might fly to Illig and an amphibious force was landed there where the fortifications were attacked and destroyed. The Dervishes were defeated and lost some 56 killed. The Isa Mahmud tribe were then invited to reoccupy the village which had belonged to them before it had been seized by the Mullah.

In spite of intensive pursuit the Mullah once again proved to be uncatchable and in July the force was broken up.

Egerton's casualties were 3 officers and 31 other ranks killed, and 10 officers and 54 other ranks wounded. He estimated he had killed 2,000 Dervishes and captured 500 rifles, 24,000 camels and 36,000 other stock. The expedition cost £2,500,000.

These campaigns were hard on the transport animals. For instance, of 4,731 camels employed on the Obbia column, 3,064 died, and the survivors were worn-out. Transport for General Egerton's operation included 2,847 Indian camels, which could each carry between 240–400 lbs., 80 Cape Buck Wagon, 482 Ekkas, 1,973 Mules and some 7,000 camels purchased locally. Of the Indian camels, 1,324 did by March 1904 and more than half of the Somali camels were also lost. The wagons and ekkas were of no use, nor were the mules.

During the 3rd and 4th Expeditions the scale of rations for British personnel included one bottle of Worcester Sauce per man for 10 days as an anti-scurvy precaution.

In October 1904, with the agreement of the British, the Italians negotiated with Haji Muhammed Abdullah, who had crossed within their frontiers, and in March 1905 the Mullah entered into a treaty, accepted the protection of the Italian flag and settled not far from Illig, on the Mijertein coast.

P.R.O.'s Medal Roll File ADM 171/56 shows 3,733 Royal Navy personnel from H.M. ships Cossack, Dryad, Fox, Harrier, Highfligher, Hussar, Hyacinth, Merlin, Mohawk, Naiad, Perseus, Pomone, Porpoise, Redbreast entitled to this clasp.

There are five volumes showing Army Personnel who were entitled and the units involved included, apart from the orthodox infantry battalion, 28 Mtn. Bty., Bikanir Camel Corps, Baluch Camel Corps, Coast Camel Corps, 54 & 55 Silladar Camel Corps, 56 S & T Camel Corps, Arab Police Corps, Somaliland Burger Contingent, R.I.M.S. Hospital Ship Hardinge, 18th & 20th Army Bearer Corps, Punjab Police, Survey of India, 15 Pack Mule Corps, Indian Pioneer Corps.

Details from the whole Indian Army list include 2 Lancers, 3 Skinner's Horse, 4 Cavalry, 6 P.W.O. Cavalry, 7 Lancers, 9 Hodson's Horse and the Army Bearer's Corps. The Army Bearer's Corps no longer existed in the Hitler War and the imagination boggles at them on parade—to what commands did they respond? 'Koi Kai' or what other loud cry. This unit did not however consist of servants for the officers but provided bearers for the regimental field ambulances. Their personnel were mainly recruited from a class of hereditary carriers, the Hindu Kahars from Oudh.

There was no resident Italian Administration in the Illig area, which was still controlled from the Italian Consulate in Aden, so the Mullah was therefore left much to his own devices and continued his intrigues with the Dolbahanta tribe. Reports reached Berbera that his agents were seeking to subvert the tribes in British territory and that he would return once again to wreak revenge on all who did not support him and to throw off the British yoke. The pressure on the Mullah's followers through over-grazing in his reserves in Italian territories gave colour to the stories that he would move back into the British Protectorate.

At this time the number of troops in Somaliland had been allowed to run down and now consisted only of the 6 K.A.R., which had 4 Somali and 2 Indian companies totalling 650 rifles. The tribal militia had been disbanded and many forts and blockhouses destroyed or abandoned.

The 6 K.A.R. was quite insufficient to defend the territory, let alone mount an expedition against the Mullah's expected invasion from Italian Somaliland and it was decided to reinforce Somaliland with 1,500 troops from East Africa and Aden, of which 1,200 were from the K.A.R.

SOMALILAND 1908–10

Army Order No. 29 of 1911 approved the award of the African General Service Medal (as instituted by his late Majesty King Edward VII) with a clasp inscribed 'Somaliland 1908–10' being granted to those officers, soldiers, and others of the Imperial and Colonial Forces who, under the command of Colonel J. E. Gough V.C., C.M.G., Major (temporary Lieut.-Colonel) J. A. Hannington, Indian Army, or Lieut.-Colonel F. J. Fowler, D.S.O., Indian Army, took part in the above-named military operations between the 19th August, 1908 and the 31st January, 1910 (both dates inclusive) provided their claims are approved by the Army Council.
The medal and clasp will also be granted in silver or bronze as the case may be to all authorised Government followers who accompanied the troops so engaged.

There was no fighting except against monotony, and most, if not all, of the troops eventually returned to their home stations without seeing a Dervish, let alone firing a shot, but they did receive a medal!

2,351 personnel from Her Majesty's ships were entitled to this clasp as under:—

Barham	25th March–12th May, 1909
Diana	7th February–22nd March, 1909
Fox	18th September 1908–1st February, 1909
Hyacinth	14th–23rd October, 1909
Philomel	19th January–20th June, 1909
	30th September–23rd October, 1909
Prosperine	1st–7th September, 1909
	14th October, 1908–23rd January, 1909

His Majesty's Government sent instructions to Berbera to abandon the interior and in April 1910 there began a two year period of 'Coastal Concentration.' This policy was a disastrous mistake as with cheap rifles available, raiding and fighting spread rapidly through the Protectorate and eventually caravans dared not leave the coast. The Mullah took this opportunity of returning to Dolbahanta country and it is reckoned that during these two years of chaos in the interior the Mullah and his Dervishes killed off one third of the population.

In 1912 the authorities sanctioned the formation of a quasi civil Camel Constabulary of 150 commanded by Richard Corfield, which moved into the interior and did manage to pacify the Mandera area.

In 1913 the Mullah built vast forts in the Nogal Valley at Tale. These were built with the help of Arab masons and were substantial affairs with walls 12'–14' thick at the base and towers 60' high. The Camel Constabulary were not allowed to approach this 'no go' area of British Protected Territory.

The Mullah's raiders now penetrated as far as Burao and in August 1913, Corfield and 300 Dolbahanta tried to recapture some stolen stock. The raiders counter-attacked, the Dolbahanta fled, Corfield was killed and if the Dervishes had not run out of ammunition, probably all the Constabulary would have been killed too. Thereafter Burao was evacuated and the Dervishes followed up and sacked the post.

In 1914 the Camel Constabulary was reorganised in military lines and styled 'The Somaliland Camel Corps' with 18 British Officers and 500 Somali rank and file, two companies mounted on camels, the third on ponies.

SHIMBER BERRIS 1914–15

The African General Service medal had already been issued with the head of Edward VII during King George V's reign (West Africa 1908 and 1909–10).

Army Order 89 of 1916 approved a new medal being struck to commemorate local military operations against native tribes or rebels in East, Central and West Africa.

The medal will be known as the 'Africa General Service Medal' and will be similar in design to the Africa General Service Medal instituted under Army Order 132 of 1902 but will bear on the obverse the effigy of His Majesty King George V.

The ribbon will be the same as that of the Africa General Service Medal instituted in 1902.

Army Order No. 90 of 1916 authorised the clasp 'Shimber Berris 1914–15' for services in Somaliland 1914–15 to officers, soldiers and others in military operations under command of Lieut. Colonel T. A. Cubitt, D.S.O., Royal Artillery, who took part in the operations against the Dervishes at Shimber Berris, between the 19th and 25th November 1914, and between the 2nd and 9th February 1915, both dates inclusive.

In July Lt. Col. T. Ashley Cubitt, D.S.O., (R.F.A.) was appointed O.C. Troops and Deputy Commissioner with Major G. H. Summers (26 Lt. Cav.) as Staff Officer.

During 1914, headquarters were set up at Burao. The troops available consisted of 150 rifles 73 Carnatic Infantry, 150 rifles of the Camel Corps and 150 Indian Mounted Infantry.

The Mullah, with the help of a German technician and Yemeni masons, had built a series of masonry blockhouses on the plateau of the Burdab range, each capable of holding fifty men. The forts had walls 12' thick at the base. Fields of fire had been cleared around the forts and the area was full of caves.

On 17th November 1914 Cubitt marched out of Burao with 14 officers and 520 Indian and Somali other ranks and arrived to within three miles of Shimber Berris before he was discovered. One of the forts was rushed and captured but the second fort was more firmly defended and could not be taken. Casualties were suffered including Captain A. Carton de Wiart (4 D.G.) who was wounded twice (he later became a General and a V.C.). In his autobiography he describes how throughout the action the enemy kept up a running fire of insults and he was so near the Dervishes that he could touch their rifles with his cane.

The force then withdrew and returned on the 23rd November with an antiquated 7–pdr. gun. This had little or no effect on the walls, but it was too much for the nerves of the Dervishes, who fled. The forts could not be destroyed as there was no high explosive available and immediately afterwards they were reoccupied by the Dervishes. The cost of the initial attack was 1 officer and 5 men killed, and 2 officers and 25 men wounded.

In February Cubitt was back with 15 officers, 570 other ranks, 6 maxim guns 2 7–pdrs. and a detachment from the 23 Pioneers with a supply of high explosive.

The forts on the ridge were found to be unmanned and were destroyed by the Pioneers. The force then moved to attack the forts in the ravine which were taken and also destroyed, while those Dervishes who were sniping from the caves were bombed out. This time the casualties were less; 5 killed and 10 wounded.

The Medal Roll consists of 77 loose pages

Military H.Q.	17
Political & Civil	11
All ranks Indian Contingent K.A.R.	306
23 Sikh Pioneers	30
Somali Camel Corps	457
Total	821

In 1937 I met General Cubitt at the Oakfield Club, Newmarket. On the wall was a chain mail shirt. 'You see that, my boy—I shot it.' Sure enough, on a subsequent inspection, there were the bullet holes. I believe this trophy came from Omdurhman.

Throughout the next five years the Mullah continued to raid into British territory and several small actions were fought with the Somali Camel Corps, but no medals were awarded, though participants were entitled to the British War Medal 1914–20.

The tempo of the Mullah's raids was stepped up. There were two not inconsiderable actions. In October 1912 at the Endow Pass and in February 1919 at the Ok Pass. The Dervishes were severely handled at both and suffered very considerable losses. These were the first actions of note of the Somaliland Camel Corps which had replaced the Corfield Constabulary in 1914.

In 1916 the Navy joined in with a sea to land shoot. Some 2,000 Dervishes swooped down from the mountains to Jidali and surrounded the coast town of Las Khorai, occupied by the Wassangeli tribe. The western quarter was captured and some 300 women and children massacred. The rest of the town offered a brave resistance for the next four days. When the first shot was fired, some seafaring Warrangeli dashed into the sea, seized a dhow and sailed for Aden. By great good luck, the wind was favourable and they made Aden in two days. Often, the voyage would take a fortnight. H.M.S. Northbrook sailed at once and arrived off Las Khorai the next day. On seeing her, the Dervishes, who had failed to capture the town, beat a hasty retreat along the coast in a westerly direction, towards the mountain pass leading to Jidali. The Northbrook kept pace with them off shore. What follows is described by Conrad Cato in 'The Navy Elsewhere, an Outpost of the Empire' (Constable 1919).

'The range was just over 6,000 yards, the shell was lyddite. After the first round, the compact group was compact no longer, but small groups showed themselves from time to time, each one of them receiving its dose of high explosive, until finally the Pass was reached. Here the Northbrook hove-to and, as the rabble came together in that narrow neck, whose range was calculated to a nicety, so the shell dropped plumb on top of them. It was diabolically easy— so easy that the sporting instinct must have sickened from it, even as our men at Omdurman sickened at the sight of the havoc their machine guns created. After 24 rounds the Northbrook ceased fire. The number of their victims will never be known, for many dead and wounded were carried off by their comrades; the only available evidence of the result of these five minutes of gunnery was afforded by the 171 corpses lying at the entrance of the pass.'

With the closer policing and administration of the Protectorate it had become clear to the Government that the Mullah's following and influence had greatly waned and the Dervish power was much debilitated. The Governor was able to represent to Downing Street that operations on a very modest scale would suffice to overthrow the Dervish power and, in October, 1919, the Cabinet sanctioned the necessary aerial and military operations.

The Mullah, who was now suffering from barasheh, caused by a dietary deficiency, had become enormously fat and it took six men to lift him onto his horse. His conduct too became more and more ruthless, which was partly a symptom of his illness.

SOMALILAND 1920

Army Order No. 170 of 1921 authorised the African General Service Medal with clasp 'Somaliland 1920' being granted to those officers, soldiers and others of the Imperial, Indian and Colonial Forces who under the general directions of the Governor and Commander-in-Chief of the Protectorate, Sir G. R. Archer, K.C.M.G., and others under the command of Major (temporary Colonel) G. H. Summers, C.M.G. (26th K.G.O.) Light Cavalry, Indian Army, and Officer Commanding Troops, Somaliland, took part in the abovementioned military operations between the 21st January 1920 and the 11th February 1920.

It should be noted that the suspender of this medal does not swivel.

The Forces deployed consisted of:—

R.A.F.:—
36 officers and 189 other ranks with a flight of DH9 bombers under command of Group Captain R. Gordon.

A Force:—
Col. G. H. Summers, 700 rifles including the Camel Corps, 101 Bombay Grenadiers, 300 Illalos and a portable wireless set.

B Force:—
Various K.A.R. contingents.

The Navy was represented by Her Majesty's ships Odin, Clio and Ark Royal.

'D' Day was 21st January 1920 and the campaign was to be opened by a bombing attack on the Mullah's forts. The crews of all but one plane lost their way, but this one machine bombed Medishe and nearly killed the Mullah. Four other planes also bombed Jidali and further air raids were made in the next few days.

The attack on Baran Fort was mounted from Las Khorai. The fort was a square building with corner towers 40' high. It was very solidly constructed and was defended by about 100 men. The fort was mortared but it was decided that it was insufficiently damaged to be attacked, which as there were only six direct hits out of 321 rounds fired, is hardly surprising. The attack was resumed on the 23rd at a shorter range with more effect and in the evening a demolition party exploded 100 lbs. of gun cotton against one of the towers. On attacking at first light the following morning, the fort was found to be unoccupied. The destruction of Baran and the other forts then took only a few days.

In this last and final round against the Mullah, the British had the advantage of communication by wireless and the use of aircraft.

One by one the Mullah's forts and strongholds were captured and destroyed and even the maxim guns lost at Erigo and Gumburu were recovered. Nevertheless, the Mullah still managed to elude his pursuers. He fled to his fort at Tale, a gigantic fortification which dwarfed any other in Somaliland. Lt. Col. H. L. Ismay (21 Cav.) set out with the Camel Corps in pursuit and chased him southwards to the border when he crossed the Haud to Galadi. In the process most of his sons and relations were either killed, captured or surrendered, but the Mullah fled on into the Ogaden country between Wal Wal and the Weli Shebah. Here he was visited by a mission of some influential sheiks who found him quite insane.

The Mullah died of influenza at the age of 56 in November 1920, and thus ended a man who for twenty years had fought great odds and managed to elude all the forces which the British and Ethiopians could deploy against him. Nowadays the Mullah is known in Somalia as Sayyid Muhammad Abdille Hassan and is revered as a great nationalist who fought for freedom.

British casualties in this campaign were trifling.

The Medal Roll consists of 230 loose pages and does not include claims by the R.A.F. and the R.N. and Royal Indian Marine, which are presumably with the respective medal offices.

H.E. The Governor, H.Q. and Intelligence	75
2 K.A.R.	228
6 K.A.R.	659

Somaliland Camel Corps	609
Somaliland Police	61
Indian Contingent Somaliland Camel Corps	140
Police attached Camel Corps	112
1/101 Grenadiers	502
Medical, P & T and S & T	46
Tribal Levy	45
Total	**2477**

Capt. A. Bigg, D.S.O., D.C.M. received the Medal and all 6 Somaliland clasps.

CHAPTER FOURTEEN

Jubaland Campaigns

HISTORICAL BACKGROUND

The medals and clasps awarded for operations against the Somal tribes
of Jubaland were as follows:—
East & West Africa Medal, clasp 'Juba River 1893'
East & Central Africa Medal, clasp '1898'
African General Service Medal, clasps, 'Jubaland,' 'East Africa
1913–14,' 'Jubaland 1917–18'

Jubaland is now part of Somalia and lies about 150 miles to the north-
east of Lamu. The Juba River formerly formed the boundary between
the British and Italian spheres of influence. The mouth of the Juba River
is closed by a bar with a depth of only a fathom at high water, but
inside the bar there is a fine navigable waterway which is 250 yards
wide at Yonte, and in 1864 a steamer with a 2½ foot draught was taken
up to the rapids at Bardera, a distance of 150 miles from the sea, where
it was wrecked in the rapids above the town. This shallow draft
steamship belonged to the German explorer, Herr Von Der Decken.
Generally, Jubaland is a dry and arid country covered with bushes.
In certain areas there are lakes or swamps which are dependant for
water on the overflow from the river.
I.B.E.A.Co. took over the administration of the country in 1891, and
the Sub. Commissioner's headquarters were located at Kismayu, some
12 miles southwest of the Juba River mouth. Kismayu has an all-weather
harbour.
The hinterland was sparsely inhabited by various Somal tribes, and
as in most desert countries, tactics were dependant on the water supply.
A strip of forest exists for 60 miles to the west of the river and the
Gosha lakes, which dry up except in the rains. The inhabitants of this
area are called the WaGosha and are mostly runaway slaves and Swahili.
The Marehan Somalis live on the right bank of the river above Bardera
and the Ogaden Somalis near Lake Deshek Wama. The Herti tribes
live at the coast, whilst far in the interior lived the Boran Galla.

THE JUBALAND CAMPAIGNS

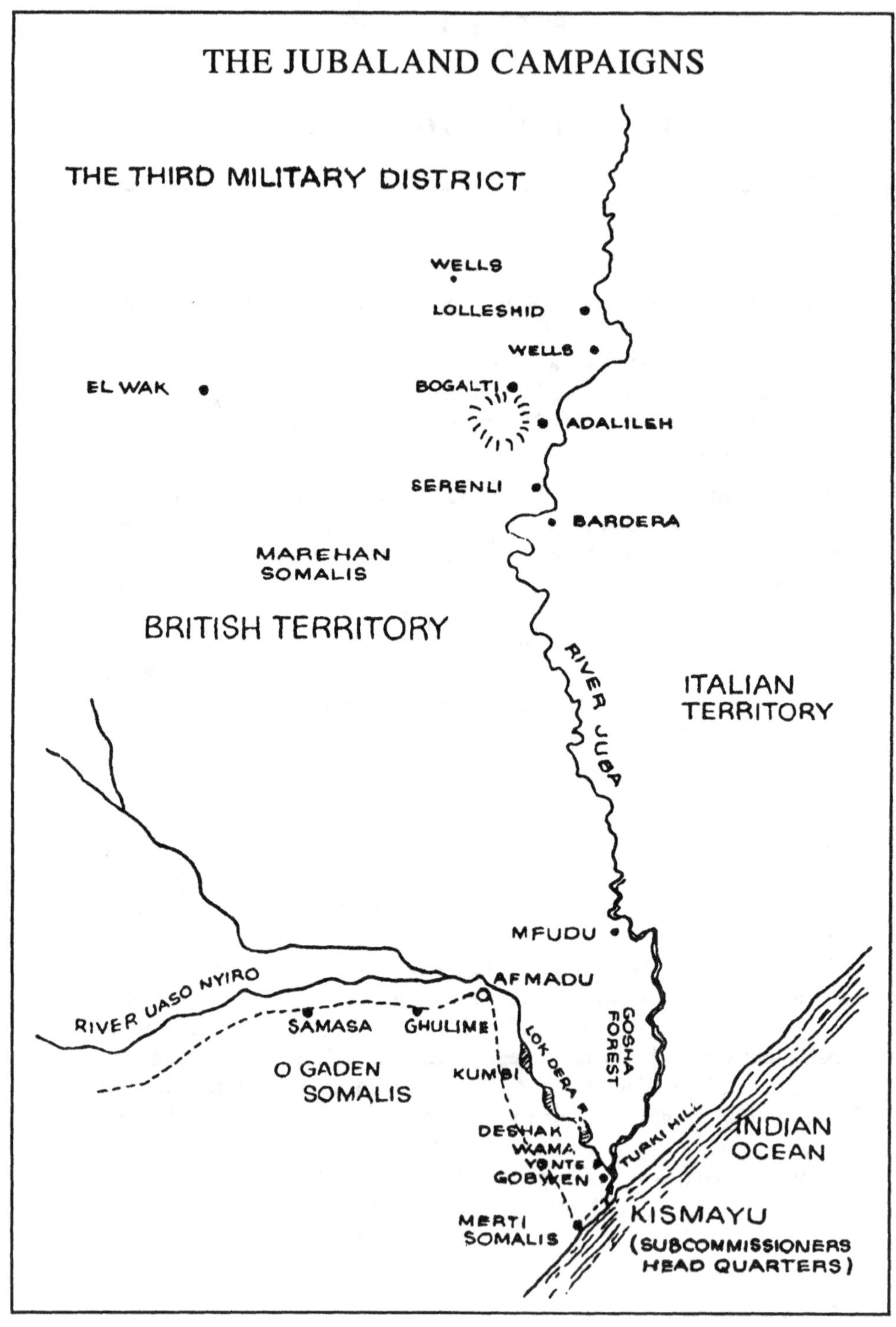

East & West Africa Medal

JUBA RIVER 1893

(23rd–25th August 1893)

This is the first of the five medals awarded for expeditions against the Somals of Jubaland.

The authority for the award of this clasp is given in London Gazette No. 26657 of 1895 (for details see Witu 1890). Full details of the action are given in London Gazette No. 26466 of 12th December, 1893.

The Sultan of Zanzibar's troops, who were stationed at Kismayu, were greatly disliked by the Herti Somals and were withdrawn by the Administrator Mr. Todd, to try to appease the local tribe.

This did not work however, as the young tribesmen became increasingly truculent and Todd was attacked and wounded at a baraza (meeting) in February 1893. He was later rescued by a landing party from H.M.S. Widgeon (Lt. W. J. Scullard) and in the subsequent skirmish some Somals were killed.

Reinforcements were sent under W. G. Hamilton, designated as Superintendent of Askaris. He had served in the Franco–German War and had a reputation of being a Prussian Martinet. He built a fort beside the river at Turki Hill close to Gobwen as fresh water supplies were available at this site, whereas the water at Kismayu was brackish. But at this stage the fort merely consisted of a newly planted thorn fence.

The garrison consisted of 46 I.B.E.A.Co.'s regular troops and 423 irregulars or Viroboto some 54 of which deserted with their arms. The deserting levies were known as the Hyderabad Contingent as they had served for a time in India.

The deserters joined the Somals and later attacked Turki Hill shooting Hamilton through the heart.

They then attacked the Residency at Kismayu where they were driven back by Count Lovatelli and Mr. Farrant, the I.B.E.A.Co.'s representative, with the small force at their disposal.

The 3rd class cruiser, H.M.S. Blanche (Lt. Price Vaughan Lewes) then arrived from Zanzibar. He landed at the river's mouth where he was joined by 50 loyal keribotos. They marched by night and at 1.00 a.m. retook Fort Turki Hill.

He then rescued two Englishmen, Capt. Tritton and Mr. McDougall from the Company's shallow draft stern wheel steamer Kenia, which had broken down in the Juba River at Gobwen.

The Kenia had been built by Kincaid & Co., of Greenock. She was 80′ x 21′ with a draft of 18″ light and 39″ fully loaded. She was driven by a wood-fired locomotive boiler and designed to give a speed of 10 knots.

She had been built to operate on the Tana River but this proved an unsuitable waterway. She was equipped with 'a novel means of defence against native canoes in the shape of a perforated tube which ran round the vessel underneath the gunwale, connected by a pipe to the main boiler and a cloud of steam could be made to envelope the steamer at will. A more practical defence in the form of a Q. F. Hotchkiss gun was fitted forward on the promenade deck.

Having repaired the donkey feed pipe, Lt. Lewes put the ship in a state of defence by means of breastworks of iron plates, cut-up canoes, sand bags and bales of piece goods, with two maxim and two Hotchkiss guns mounted for armament.

They proceeded up river and shelled and destroyed Magarada and Majawen, where the Somals armed with sniders replied to the Kenia's fire, but they were defeated and fled into the bush.

The naval contingent then returned to their ships leaving the Kenia moored by three anchors in the middle of the Juba River, in the charge of two friendly chiefs.

The insurgent Herti Somals then vanished into the bush and for some time there was no further contact with them.

Honours and awards given were:—

Capt. G. R. Lindley	C.B.
Lt. Price Vaughan Lewes	D.S.O.
Count Giovanni Lovatelli, Cavalleri della Corona d'Italia	Honorary C.M.G. 3rd class

Adm. Roll 171/46 also shows that 45 persons were entitles to this clasp including 5 seedies—one medal to a seaman was returned 'D.D.'

H.M.S. Blanche. Juba River 1893

Lt. E. V. Lewis	Mr. L. Barber Gunner
Shipwright J. Avery	A.B. W. Bellamy
C.P.O. J. Bennett	E.R.A. 3 Cl G. J. Carey
Ch. Writer A. Betts	A.B. C. Grant
A.B. J. Denham	S.B.S. A. Johnson
2 Yeo. Sig. T. Hardings	A.B. C. Lyne
Cooper J. J. Millet	Sto. H. Palmer
P.O. 1 CL H. Smith	Ord. H. Singer
L/Sea. S. Train	Sto. A. S. White
Ch. Arm. W. Henwood	

Seedies Yussif Bin Abdullah and Feroze Sumaheli

JUBA RIVER 1893 & WITU 1893

A.B. T. J. Bridle	L/Sea. H. Grant
A.B. G. A. Clark	Also Brass River 1895
Sto. G. T. Cudlip	L/Sea. T. J. Cole
A.B. C. F. Evans	A.B. C. Clift
Sto. W. Edmonds	Ord. A. Davey
A.B. A. Howard	Sto. W. Easterbrook
L/Sea. E. S. Maddoch	A.B. G. R. Locke
2/Cpl. H. C. Marsh	A.B. A. Maddock
P.O. 2 Cl W. H. Potter	Sick Bay At. J. Pedley
Cpl. E. Reading	P.O. 2 Cl W. H. Taylor
A.B. J. Weeks	A.B. W. J. Wade
A.B. E. Richardson	

Krooman Moose Bin Assou, Othman, Jinnah Maychone

Whilst not on the roll, a medal was issued to K. MacDougall Dist. Officer. It is correctly impressed and has 1895–6 for Mwele. He also was awarded 1898 East & Central Africa, Jubaland and the I.B.E.A. Co.'s Medal. As he was rescued, he was obviously entitled to the medal as was Capt. Tritton.

For the Admiralty Despatch see London Gazette dated 12.12.1893, p. 7254 et seq.

EAST & CENTRAL AFRICA MEDAL 1897–98

1898

Army Order No. 125 of 1899 authorised the East & Central Medal to the forces engaged against the Ogaden Somalis from April to August 1898. The clasp to be inscribed '1898.' The medal was awarded in silver and bronze.

The Ogaden Somalis, who live near to Lake Wama, are a group of fierce and warlike tribesmen. Their chief, Sultan Ahmad Bin Marghan was estimated to be able to muster 6,000 warriors. The Ogaden possessed few fire arms and were armed with spears, knives and small round shields made of giraffe hide. Their favourite tactics were ambushes and fighting in thick bush, where they generally fought in pairs, one warrior seizing his enemy whilst the other speared him.

There were the usual preliminaries to a punitive campaign; cattle thefts, murders of 'friendlies,' policemen and finally, following police action, the Somalis raided towards Kismayu killing two Arab traders.

The expedition was commanded by Major W. Quentin (4th Bombay Rifles) who had arrived in Mombasa from India in March with a wing of his regiment. The force included:—

4th Bombay Rifles	wing
27 Baluchis Bombay Infantry	2 coys
East African Rifles	2 coys
Uganda Rifles	4 coys
Totalling	1,060 all ranks

The force reached Turki Hill in April and found that the police post at Yonte, which is situated about ten miles north of Gobwen, had been captured and fifteen police killed. By May the main force was based at Yonte, which had by then been stockaded. The Ogaden attacked the cattle boma outside the Fort, but were driven off with losses.

An advanced base was later established at Helished near Lake Wama, about a day's march from Yonte.

On June 30th an Indian patrol from Yonte of forty-two all ranks was ambushed and twenty-seven were killed including their officer. At this stage initiative seems to have been with the Ogaden, and in July they even raided cattle belonging to the Government at Kismayu but reinforcements then arrived at the port and released all of Major Quentin's forces for offensive operations.

On 3rd August 300 Rifles attacked and surprised the Ogaden camp near Helished, inflicted heavy casualties and captured 450 head of stock. A few days later, further cattle were taken.

This was sufficient to make Ahmad Bin Marghan and his chiefs sue for peace and in consideration of a fine of 500 cattle and the return of the rifles captured in the earlier fighting, peace was agreed.

The cattle fine was paid grudgingly and only the worst and oldest of animals were produced.

W.O. File 100/90 shows the following were entitled to the medal and clasp:—

1st Bn. Baluch Bombay Inf.	163
46 Native Field Hospital	9
4 Bombay Rifles	777
Native Field Hospital	17
1st Bn. Uganda Rifles I.O.R.s	375
Followers attached thereto	45
	1,386

An I.O.M. was awarded to Nk. Butta Singh (4 Bom. Rifs.) for gallantry near Lake Wama on 22.6.1898.

African General Service Medal

JUBALAND

Army Order No. 133 of 1902 authorised the medal with clasp 'Jubaland' to all officers and men who took part in the operation against the Ogaden Somalis under the command of Colonel T. Ternan, including the military forces at Kismayu, and to such Officers and men of the Royal Navy and Royal Marines serving on His Majesty's ships 'Magicienne,' 'Terpsichore' and 'Scout,' as were landed between the 16th November 1900 and 30th April 1901.

After the 1898 campaign, when the punishment inflicted on the Ogaden was not sufficiently condign, a garrison was established at Yonte.

The Sub Commissioner, Mr. A. C. W. Jenner, who met with a generally friendly reception when he toured his district, believed that the Ogaden had the highest qualities, and was generally misled and unaware of the revengeful feelings harboured by some of the tribesmen.

In November 1900 Hasan Yere, who bore Jenner a grudge for having imprisoned him, attacked his sleeping camp in the early morning, killing Jenner and his escort of 9 Hati Policemen, whose rifles were not even loaded. Other murders then occurred and the inevitable punitive expedition was sent out under command of Brigadier General T. Ternan (Manchester Regiment), who with Lt. Col. G. P. Hatch brought reinforcements from Nairobi and Mombasa.

Qmar Bin Margham, the Sultan's brother, who was suspected of complicity in Jenner's murder, was summoned to Kismayu, but he did not answer the summons.

Hatch, being sick, was replaced in command by Major T. Souter (Cam. High). The force at Kismayu consisted of 5 coys E.A.R. (9 officers and 475 O.R.S), 8 officers and a wing of 16 Bombay Infantry, 52 rifles Aden Camel Corps, a coy of Somali police, and for the first time in Africa, the Indian Artillery with a section from the 9th Murree Mountain Battery, and 900 porters. With armed porters, the force totalled 1,404 armed men and 600 camels formed their transport train.

A base was established at Yonte and in February Afmadu was occupied and the Sultan, who was generally regarded as a comparatively pacific man unable to control his adherents, surrendered and was sent by the military to Kismayu on the charge of being privy to Jenner's murder.

A strong base was set up at Afmadu and a flying column with one gun attached was sent to Samasa about 57 miles to the north-west in February. The country was unknown and turned out to be practically waterless and covered with high, thick bush. Some stock was captured.

Tracks indicated that the enemy were present in force at Samasa. The bush was thick and Col. Ternan formed square. His column had barely taken up position when the enemy attacked the rear of the square which was held by 5 coy E.A.R., who bolted. This exposed the field hospital, the square was entered and the medical officer, Lt. Col. Maitland, killed. Eventually the company commander rallied his men and after ten minutes hard fighting the Somals were driven out and withdrew.

The flying column's losses were 17 killed and 20 wounded, whilst the Somalis later admitted to have lost about 400 amongst whom were Jenner's murderers.

The force moved back to Afmadu, fighting off several attacks from the Somalis who had to be driven off by frequent volleys. The 9th Murree Mountain Battery made good practice with case shot, killing a group of a dozen or so Somalis who showed in an open space.

After the rains broke in April several mules died from the bite of the tetse fly. The Indian troops were returned to India in June and the garrisoning of the area left to the E.A.R.

The policy adopted was to leave the deserts alone and a Camel Corps was formed by the E.A.R. to give the mobility necessary to deal with the elusive Somalis.

File ADM 171/56 shows approximately 209 entitled from H.M. Ships Magicienne, Scout and Terpsichore, so with the military, something in excess of 1,600 medals should have been issued.

African General Service Medal

EAST AFRICA 1913–14

Army Order No. 222/1916 authorised the clasp 'East Africa 1913–14' to the forces employed in military operations under the late Lieut.-Colonel B. R. Graham, 3rd Bn. King's African Rifles, against the Merehan Tribe, East Africa Protectorate, between the 15th December 1913 and the 31st May 1914, both dates inclusive.

In 1910 a post to accommodate 1½ coys was established upon a small hill overlooking the Juba River at Serenli, some 150 miles from Kismayu. For six months of the year the river was navigable by steamer from Gobwen, but the post was not entirely cut off as it was only four miles from the Italian post of Bardera who were able to communicate with the outer world by wireless.

The Civil Administration's policy towards Jubaland and the Northern Frontier District (N.F.D.) was one of 'observation,' but when tribesmen became too troublesome military action had to be taken.

The Marehan's raids on the Aulyehan became so frequent that it was decided to disarm the former tribe which was believed to be able to muster 1,150 fighting men.

The Marehan resisted disarmament. An advanced base was therefore built up at Lolleshid and in December 1913 permission was given for the K.A.R. to undertake a punitive expedition under command of Lt. Col. B. R. Graham (3 K.A.R.) against the Marehan whose attitude was becoming ever increasingly hostile.

In December an attack was made by the tribesmen on a signal station at Adalileh when casualties were suffered by both sides, and this was followed by an attack on a convoy from Serenli by a force of 400 Somalis with 150 rifles at Bogatti. The K.A.R.'s behaviour was unsatisfactory, casualties were sustained and the convoy had to return to base. This defeat greatly disturbed the administration in Nairobi. Some reinforcements were sent and 70 Jibrahil from the Aulyehan were enlisted as scouts. A series of chases and minor skirmishes then followed without much apparently being achieved, but in March 1914 a camel convoy with supplies and a company of 3 K.A.R. arrived. The appearance of these reinforcements made the Marehan think again and opposition ceased. It was then apparent that the Marehan's fighting potential had been greatly exaggerated. Fines of rifles, ponies and stock were levied and paid.

The medal rolls at the Army Medal Office which are unbound show the following as entitled to this clasp:—

1 K.A.R., including M.G. Porters & Boys	305
3 K.A.R.	301
4 K.A.R.	133
53 Silladar Camel Corps, an Indian unit who presumably provided their own camels. The roll is signed by G. E. M. Hogg, at Jullundur, as Inspector of Camel Corp	250
Serenli Military Transport Company, including the rank of 'armed syce'	114
East African Police	9
R. E. Salkeld, the Commissioners and staff (civil)	4
Col. Graham and H.Q. Staff north of Serenli	25
	1141

AFRICAN GENERAL SERVICE MEDAL

JUBALAND 1917–18

Army Order No. 95 of 1919 authorised the clasp 'Jubaland 1917–18' to all officers and men who served under the command of Lieut.-Colonel W. E. H. Barrett, King's African Rifles; Major F. G. M. Porcelli, Duke of Cornwall's Light Infantry; Captain J. F. Wolseley-Bourne, or Captain O. Martin, King's African Rifles, in the military operations against the Northern Aulihan tribe west of the Juba River and north or east of a line Waregta–Lak Abaleni–Lorian Swamp–Eil Wak–Dolo, between the 23rd July, 1917 and the 24th March, 1918.

During the Kaiser's War, with the withdrawal of most of the regular troops from Jubaland, there was much tribal unrest.

An armed constabulary, mainly mounted on camels, was formed to keep the peace, but the Aulyehan and Marehan tribes continued to carry out raid and counter-raid. In 1916 Abdurraham Mursaal, the leader of the Aulyehan, had become so arrogant that he refused the arbitration on a dispute by Lieut. Elliott, the officer commanding the Constabulary at Serenli. Finally, when Abdurraham was threatened with arrest if he did not pay blood money, he collected a force 1500 tribesmen from the Hellished area, and attacked Elliott's camp.

He had previously tricked Elliott into believing that trouble would occur if the Constabulary carried their rifles outside the boma when off duty at night. Accordingly, most of the police force's rifles were kept locked in the guardroom and when the attack came, many of the police were unarmed.

Elliott and 35 of his police askaris were killed, 7 were wounded and 3 deserted. 50 civilians were also murdered and a machine gun, 50 rifles and 12,000 or so rounds of S.A. ammunition captured.

As there were no troops available, nothing could be done to avenge this outrage for a year.

In August 1917 Lieut.-Colonel W. E. H. Barrett (Res. Off) commanding 5 K.A.R. left Gobwen by river steamer for Northern Jubaland to re-establish pax Britannica in the area west of the Juba River. Our allies, the Italians, assisted operations by preventing the tribesmen crossing the Juba River into their territory.

The Serenli post was reoccupied, the resistance of the Aulyehan overcome and Capt. O. Martin (Spec. List) was left in command.

By then, command of the troops in Jubaland had devolved on Major E. G. M. Porcelli (D.C.L.I.) and he ordered Martin to drive the enemy south away from their water supplies, towards another force advancing from Yonte, with the object of capturing those responsible for the attack on Elliott's camp.

The Marehan tribesmen were co-operating with the Administration against their old rivals and in December a small party consisting of 15 mounted infantry, 15 askaris, 350 Marehan riflemen and 450 spearmen set off for Haras where 2,000 camels were captured. The Aulyehan immediately counter-attacked, the Marehan tribesmen ran away and all but 160 of the camels were recaptured.

When the wells at Hagagabli dried up the Aulyehan had to return to the Juba River to obtain water and there they were attacked and defeated. Eventually Abdurrahan's camp was overrun, but he escaped, though many of Elliott's effects were recaptured together with 1,233 camels and 3,000 goats. These were slaughtered and a further 600 camels were captured at Hafanli.

Thereafter patrols caught most of the chiefs. Terms were imposed, fines levied and 5,000 camels brought in, together with 450 large bore rifles including the rifles and ammunition taken at Serenli. Eight of the men responsible for Elliott's murder, including Gabodi Ali, Chief of the Rer Afgab, were hanged, whilst Abdurrhan took refuge in Ethiopia.

In July 1925 the districts of Juba River and Kismayu (33,000 square miles) were ceded to Italy and became part of Italian Somaliland.

The medal rolls are unbound and show that the following were entitled to this clasp:—

3 K.A.R. all ranks	26
5 K.A.R.	657
1/6 K.A.R.	91
Northern Frontier District Constabulary	150
The Provincial Commissioner	1
	925

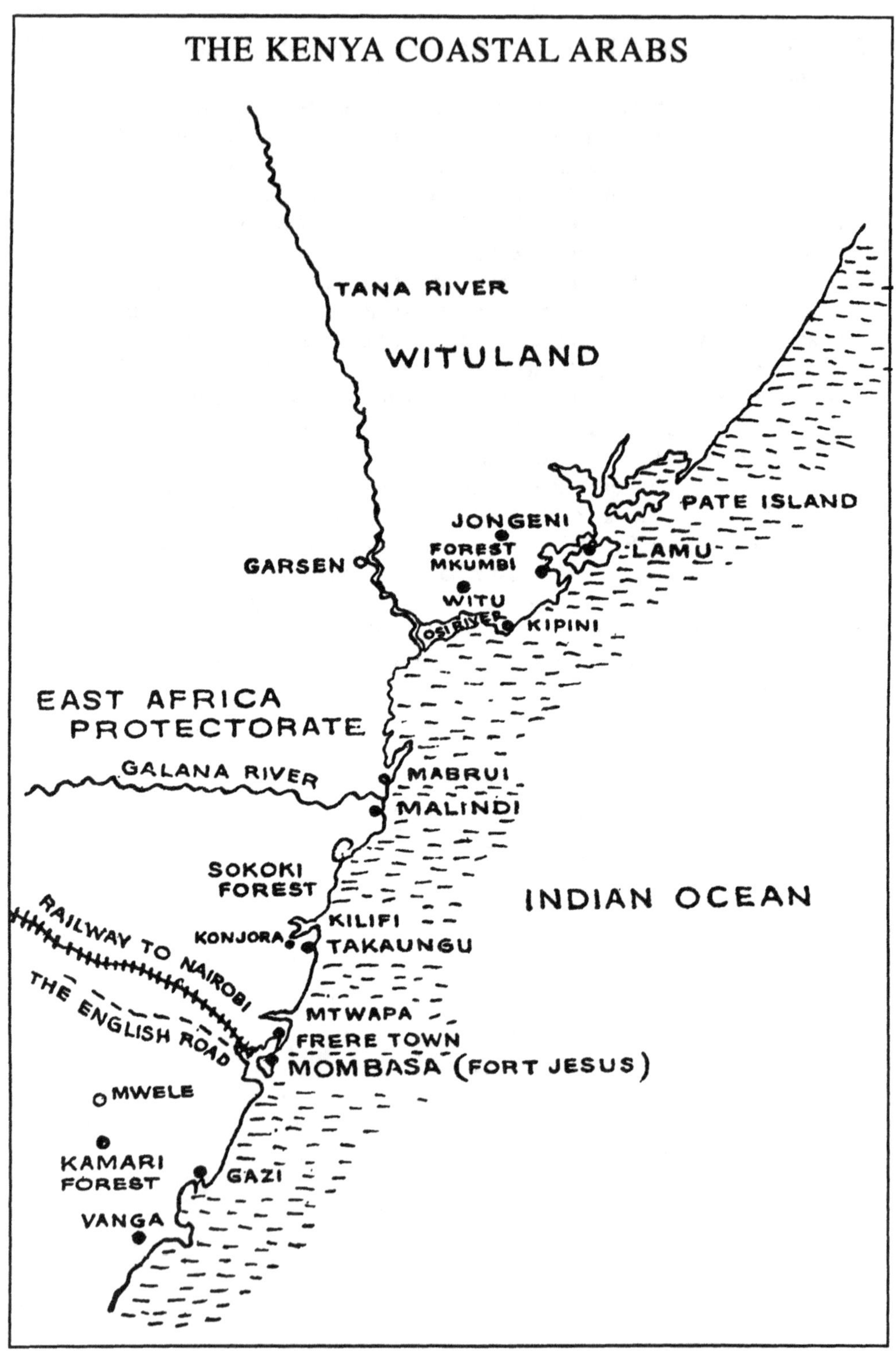

THE KENYA COASTAL ARABS
TANA RIVER
WITULAND
PATE ISLAND
JONGENI
FOREST
MKUMBI
GARSEN
LAMU
WITU
OSI RIVER
KIPINI
EAST AFRICA
PROTECTORATE
GALANA RIVER
MABRUI
MALINDI
SOKOKI
FOREST
INDIAN OCEAN
KILIFI
KONJORA
TAKAUNGU
RAILWAY TO NAIROBI
MTWAPA
FRERE TOWN
THE ENGLISH ROAD
MOMBASA (FORT JESUS)
MWELE
KAMARI
FOREST
GAZI
VANGA

CHAPTER FIFTEEN

The East African Coastal Areas

THE HISTORICAL BACKGROUND

There is very little written history about the East African Coast, where from very early times some Arab settlements had been in existence on its Eastern Sea Board.

1497   The first European contact was in 1497 when Vasco Da Gama touched in at Mombasa and Malindi. Three years later the Mombasa port was captured by the Portuguese.

1507   In 1507 Tristan Da Cunha captured Lamu.

1528   By 1528 the Portuguese Sovereignty was firmly established over the whole East African Coast south of the Equator. They built Fort Jesus at Mombasa, which still stands, and was later used as a gaol by the British.

1660   The Imam of Muscat, who held a commanding position in Western Asia attacked the Portuguese Territories and captured Fort Jesus Mombasa in 1660. By 1698 Zanzibar had been taken over and the Portuguese expelled from the East African Equatorial Coast and thereafter they only remained in the south at Mozambique.

1800   This status quo remained throughout the 18th Century. In the time of the Napoleonic Wars in 1800 Britain made a pact with the Omani Arabs at Muscat as a measure to counteract Bonaparte's designs in India and a British Resident was established at the Sultan's Court.

The preoccupation with maintaining the security of their supply line to India and the anti-slavery political lobby in the U.K. led H.M.G. to become more and more involved in the affairs of the Omani Arabs and with Zanzibar.

In the 19th Century the Littoral was part of the Omani Empire, whose main settlements were at Lamu, Mombasa and Dar Es Salaam, with many minor enclaves at the small harbours and inlets between these ports.

1824   British warships visited the East African coast and in 1824 the Mazrui Arabs, in rebellion against Oman, ceded Mombasa to the British in the person of Captain W. P. W. Owen of H.M. Ship Barracouta: this British Protectorate was, however, quickly repudiated as soon as the Honourable East India Company got to hear of it and the Union Jack was hauled down at Fort Jesus as Owen had no powers to treat.

1845 The Sultan of Oman gradually transferred his Court to the Island of Zanzibar and in 1845 another Treaty, this time to stop slavery, was forced on the Sultan by the British. For a variety of reasons it was ineffective.

A British Consulate was established there at this time.

1859 A faction war broke out between the Omanis in Muscat and Zanzibar and in 1859 a Muscat based Omani Arab invasion fleet destined to capture Zanzibar was turned back to Muscat by the Royal Navy. The Viceroy of India, Lord Canning was later

1861 called upon to adjudicate and in 1861 decreed that Zanzibar be separated from Oman.

British influence on Zanzibar was exercised by H.M.'s Consul or Balozi as he was known. His duties went far beyond those of a normal Consul and he in fact dictated H.M.G.'s policy to the Sultan.

Zanzibar was the centre of the Slave Trade and indeed all the caravans of Arab slavers operating into Central Africa invariably carried the Sultan of Zanzibar's red flag at their head.

1873 In 1873 H.M.G. really put the squeeze on and slavery was abolished in Zanzibar, though as will be noted elsewhere, it still flourished on the mainland.

1875 The then Sultan Bargash was an Anglophile who in 1875 sought to give a 66 year lease of the African Hinterland to Sir William Mackinnon, later founder of I.B.E.A.Co. This, however, came to nothing and in 1884 Carl Peters, the German explorer, backed by the German Chancellor Bismarck, slipped into the interior & signed treaties with a number of completely illiterate Chiefs. Political considerations inhibited the British from supporting their ally Bargesh who was deserted and when a German Naval Squadron arrived off his Capital he could only acquiesce to the annexation of the centre of his mainland empire. Thus German East Africa (Tanzania) was created.

1883 The U.S.A. signed a treaty with the Sultan in 1883 and later opened a Consulate in Zanzibar.

1884 In 1884 the African Lakes Company had established a trading post at Karonga on the North Western shore of Lake Nyassa.

1886 In 1886 an International Commission investigated and found that the Sultan of Zanzibar had Arab Troops at Dar es Salaam, Pangani, Tanga, Mombasa and elsewhere. It was then agreed that the Sultan's territories should be defined as a strip of coast 10 miles deep, stretching from Tungi Bay to Kipini.

The Sultan was not even represented at this commission for, as was remarked, he was merely an 'Oriental Prince to whom are not accorded the usual rights prescribed by international law.' Later, the Germans ignored this Treaty and took over the ports of Tanga, Dar es Salaam and the coastal strip. The British, however, paid lip service and rented the Sultan's Mombasa coastal territories strip.

1887   In 1887 the British East Africa Company was formed with the quite inadequate capital of £240,000 and in the next year became constituted by Royal Charter and added 'Imperial' to its title (I.B.E.A.Co.).

1890   In 1890 an Anglo German Agreement was signed demarcating the Boundaries between British East Africa (Kenya) and German East Africa (Tanzania).

1880   In 1880 the German East Africa Company proclaimed a protectorate over the area from Witu to Kismayo, The Anglo German Agreement gave this territory to Britain and gave up all treaty rights to Uganda.

1887   In 1887 I.B.E.A.Co. leased Jubaland from the Sultan of Zanzibar and in 1891 the Juba River was fixed as the boundary with the Italians.

1895   In 1895 Jubaland was taken over by the Imperial Government and was ceded to the Italian Government in 1925.

1895   In 1895 I.B.E.A.Co. gave up its efforts to administer its territories and H.M.G. acquired the Company's rights for a sum of £250,000. The Sultan of Zanzibar's coastal strip was leased for an annual payment of £11,000 but he was allowed to fly his red flag over Fort Jesus and it remained flying there until 1964 when Kenya achieved its independence.

In 1963 Zanzibar was granted independence under the Sultan but he was deposed within days and many were murdered in the fighting and riots which followed.

Campaigns against the Coastal Arabs

There were three campaigns in the 1890s against the Coastal Arabs, who for centuries had been the dominant power on the East African Coast.

EAST & WEST AFRICA MEDAL (1887–1900)

WITU 1890

(17TH–27TH OCTOBER 1890)

Full details of this expedition were given in Admiral Freemantle's dispatch published in the London Gazette, dated 6th January 1891.

In the London Gazette, No. 26657 of August 30th 1895, the Department of the Accountant General of the Navy gave notice *'that medals and clasps were now ready for issue to those officers, seamen and marines who were actually engaged and formed part of any one of the following expeditions:—*

Witu—October 1890
Chief Liwondi—February and March 1893
Pumwani and Jongeni—August 1893
Juba River—August 1893
Lake Nyassa—November 1893
Gambia—22nd February to 11th March 1894
Benin River (Chief Nanna)—August & September 1894

The medal is the same pattern as that issued for the Ashantee War and those officers and men already in possession of the medal for the former operation will only receive such of the clasps as they may now be entitled to'

Witu is situated about 25 miles inland from the mouth of the River Osi and about 50 miles west of Lamu. This area had formerly been under the control of the Sultan of Zanzibar but in 1888 the state-supported German East African Company declared a Protectorate over the coast around Witu, to Kismayu in the north.

However, the area around Witu was still ruled by Sultan Ahmad Bin Fumo Luti, a descendant of the Sultans of Patta Luti, who was nicknamed Simba (the lion). He collected 3,000 or so followers and terrorised the surrounding tribes and in 1888 when the Sultan of Zanzibar tried to send a punitive expedition against him to remove an unauthorised customs post on the River Osi, the Germans refused to allow the expedition to proceed. However a small British Force landed and occupied the district and Simba had to retire.

The Anglo–German agreement of 1890 gave control of this area to the British.

Fumo Luti had by then been succeeded by Fumo Bakari.

A German commercial syndicate had previously been granted a timber concession in the Witu Forest and in August, Herr Kuntzel, an irascible man, and ten mechanics arrived to commence operations. Kuntzel quarrelled with Fumo Bakari and when trying to escape was pursued by an infuriated rabble, armed in various ways, including bows and arrows, and was murdered. The surviving two Germans were murdered in cold blood the following day. The German Government demanded that this outrage be avenged, a task which was utterly beyond the capabilities of the I.B.E.A.Co.

Vice Admiral The Hon. Sir Edmund Freemantle, Commander in Chief of the East India Station, therefore took a squadron to Lamu from whence he sent an ultimatum to Fumo Bakari, which produced no result.

A punitive expedition then set out from the ships. The dispatch gives many details of clothing and accoutrements—officers to wear helmets stained brown—tea was recommended—brown belts, water bottles, knives, pistols and swords, blue tunics or monkey jackets with trousers and gaiters. Junior officers to wear cutlasses instead of dirks, men to wear white hats, belts and braces with water bottles, spades, haversacks, knives and blankets (with pipes and tobacco), gaiters and blue serge suits. Seedies to be provided with cutlasses, 80 rounds of ammunition to be carried, while gun ammunition was half double shell and half shrapnel—How hot they must all have been!

The British Forces consisted of between 750 and 800 sailors and marines (with 4 7–pdrs. and 4 machine guns) from H.M.'s ships Boadicea (Capt. Hon. A. G. Curzon-Howe), Turquoise (Capt. J. Brackenbury), Conquest (Capt. Henderson), Cossack (Capt. McQuhae), Brisk (Comd. Winsloe), Kingfisher (Comd. Milen Gardiner), Redbreast (Lt. Keary), Pigeon (Lt. Floyd), Humber (Lt. Brown), the hired transport Somali and I.B.E.A.Co.'s S.S. Juba. The infantry consisted of 150 of the Company's armed Indian police under Capt. A. F. Eric Smith (Life Guards), 200 of the Sultan of Zanzibar's troops and 281 Seedies attached to the ships.

On the 24th October the boats of the Boadicea, Cossack and Brisk were sent to Mkonumbi and Baltia villages where the two Germans were murdered in cold blood. These villages were taken and burnt with but slight opposition. On the next day the main forces landed at Kipini and marched on Witu.

In recent times many places in Africa have changed in appearance—forest have been cut down, dense impenetrable bush, is now open savannah, there is much cultivation, housing and even metalled highways. Admiral Freemantle's dispatch describes his route from Kipini to Witu. This route is now followed by a sand track and the country has changed very little. It still has the same heavy sand, dom palms and high grass, though perhaps the sailors of 85 years ago were not so tall as their present day counterparts as very little of the grass would be 'over men's heads' nowadays.

The advance took the expedition three days and the force occupied zaribas by night. Fumo Bakari made determined attempts to prevent the British reaching Witu. One night attack cost three wounded.

As they neared Witu, the force was again attacked, but the Witu Swahilis, half of whom were armed with firearms, were easily driven off as only a small proportion of Bakari's forces were said to have been engaged.

Witu, which was little more than a stockaded village, surrounded by dense bush, was shelled from a mound now occupied by the Kenya Police Lines. Lt. Lalor fired two rounds of 7–pdr. at the town's gate and a gun cotton party from the Boadicea then rushed forward and blew in the gate. There was little resistance in the town, and Fumo Bakari's forces were pursued for three miles and many were killed.

The town and Sultan Bakari's house, which contained a large quantity of arms and ammunition, were then burnt and the force returned to Kipini on the 28th October.

H.M.'s force's total casualties were 13 wounded.

Nothing remains of the old Witu, but the typical palm thatch, coral and daub houses there now must look much as they did then. Beside one of the old wells is Simba's tomb. The inscription reads:—

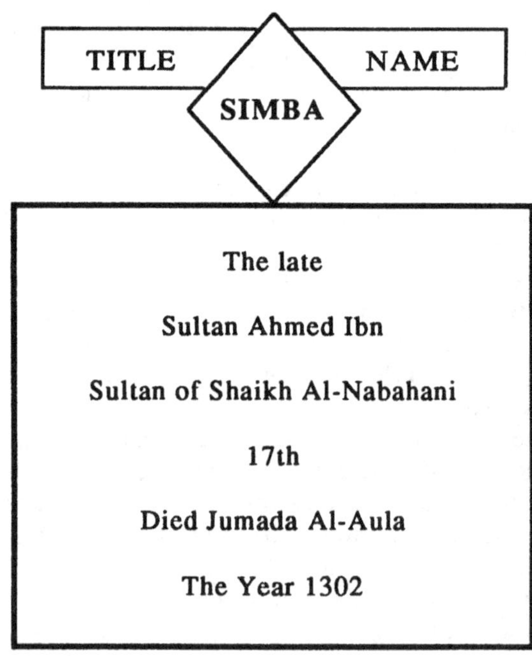

He is well remembered and seems likely as time goes on to become a type of Robin Hood of the Freedom Fighters.'

File No. ADM 171/46 at the P.R.O. further states 'By their Lordships' directions, the medal will only be awarded to those actually engaged and forming part of any of the expeditions, their Lordships' intention being to limit the grant to those who were landed or engaged in boat service in connection with the expedition, excluding those who remained on board ship.'

A note on file ADM 171/46 no medal awarded for Capt. Henderson's (Racoon) expedition of September–October 1893 vide E 960/95 & Foreign Office 5.12.93 Whitehall papers.

This description also referred to the 1893 Vitu (spelt thus) expedition, Liwonde 1893, Juba River 1893, Lake Nyassa 1893, Gambia 1894, Benin River 1894, Brass River 1895, Mwele 1895, Benin 1897 and 1897–98 (Borgu and Illah).

In the roll this clasp is referred to as the 'Ashantee Medal,' not the 'East and West Africa Medal.'

Admiral of the Fleet, Sir Roger Keyes, obtained this clasp as Midshipman R. J. B. Keyes of H.M.S. Turquoise.

The medal was awarded to personnel from the following H.M. Ships: Boadicea, Brisk, Conquest, Cossack, Humber, Kingfisher, Pigeon, Redbreast and Turquoise. The roll shows some 1,059 naval personnel received this clasp.

295 Seedies also accompanied the expedition including one Abdullah bin Hamed (Snowball).

124 of the medals awarded to naval personnel were returned to the Mint. It is not always apparent why, but in some cases the roll is marked 'D.D.' or Run on ...'

I have been unable to trace a medal roll for the Police or Zanzibari troops.

WITU AUGUST 1893

(7TH–15TH AUGUST 1893)

The authority for this clasp is in the London Gazette of 30th August 1895 (see Witu 1890), where this campaign is referred to as 'Pumwani and Jongeni August 1893.'

Full details of the expedition are given in Admiral Bedford's dispatch published in the London Gazette No. 26466 dated 12th December, 1893.

After the 1890 expedition, the I.B.E.A.Co. was responsible for the administration of the Witu area and garrisoned it with Indian police.

The Police's contracts expired in 1893 and as none of them were willing to renew their contracts, other arrangements had to be made, particularly as I.B.E.A.Co. considered that the obligation to garrison Witu should be undertaken by Her Majesty's Government. After some argument the Company withdrew its forces and with effect from the 31st July, 1893, the responsibility for law and order in the Witu district reverted to the Sultan of Zanzibar.

The Sultan of Witu at that time, Fumo Omari, Fumo Bakari's brother, was described as 'restless and dangerous and openly defiant' as he refused to meet the British Consul General when he visited Witu to proclaim the new order. Omari then began to commit outrages, though what these misdemeanours were, history does not relate,but it seems most likely that the authorities feared the Omari's Roga Roga would overrun the district unless the Indian police were replaced.

Omari's two strongholds were Pumwani and Jongeni, but these places no longer exist, even as names on the maps. It is often difficult to locate villages mentioned in dispatches as apart from the fact that many such villages were destroyed as a result of the punitive expedition. Some of these villages were slave camps set up to save the slaves a long walk back to their Arab owners' homes after working in the plantations. Mkumbi also no longer exists, but Omar Atik, a descendant of the Sultan, told me that these villages were all near Witu. Another place, Mambo Ya Sasa, meaning in Swahili 'Act now,' appears on the map. It is where Simba and his followers met to decide what to do next. There is nothing there now except some very old mangoe trees, perhaps from the seeds dropped by Simba's followers.

To assist Mr. Rodd in declaring a Protectorate over the territory of Witu and to punish Omari, a naval force commanded by Capt. G. R. Lindley (3rd class cruiser Blanche) was sent to the area. This consisted of H.M. Sloop Swallow (Com. L. Dod Sampson) and the gunboat Sparrow, and landed a naval brigade of four companies of seamen, with some kroomen and seedies, a company of marines, 40 Sudanese and 20 Zanzibaris under command of Capt. G. P. Hatch (Wilts. Regiment), who subsequently commanded the 3 K.A.R.

Pumwani and Jongeni, which had defied the I.B.E.A.Co.'s police, were attacked. The gates of Pumwani were blown in by a field gun and war rockets, and the towns destroyed. Although the fighting was described as 'brisk,' the casualties suffered by the naval force were only one stoker killed and two officers and six seamen wounded.

A further naval brigade known as the Lamu Forest Expedition was therefore sent in October from H.M.S. Racoon (Comd. F. M. Henderson), H.M.S. Blanche and H.M.S. Swallow. It was accompanied by some Zanzibari troops. Pumwani was taken and destroyed. Casualties this time were 3 wounded, but there were many cases of fever.

Omari was deposed and in 1895 Omar Bin Hamid was installed as ruler of Witu and thereafter there was tranquillity on this portion of the coast.

It is interesting to note that when I was inquiring from an old Arab shopkeeper at Malindi whether any descendants of the Sultan's remained, I was asked, 'Did I mean the real Sultan or the Usurper?' Memory of Simba lives on and it is predictable that he will become one of Kenya's folk lore heroes for the fight for freedom.

Enemy firearms and weapons captured are listed in the dispatch to include: 30 repeaters, 100 breech loaders, 20 carbines, the rest cap, etc. etc., also bows and arrows (poisoned).

According to Gordon, medals were awarded to 70 natives of the West India Regiment and Niger Constabulary, and 266 to the Naval Brigade.

File No. ADM 171/46 refers to this expedition as 'Pumwani and Jongeni.' Some 286 Royal Naval personnel and 36 seedies, kroomen and tindals were entitled to this clasp.

Obviously there must be some roll somewhere in which the West African personnel are included, though I have been unable to find out what these men can have been doing so far from home, and can only conjecture that they were aboard H.M. ships, if in fact troops from West Africa did participate in this campaign.

MWELE 1895–1896

In the London Gazette No. 26816 of 22nd January 1897 the Accountant General of the Navy gave notice that 'the medals for the Mwele expedition August 1895 were now ready for issue to those officers, seamen and marines who were landed from Her Majesty's ships, St. George, Phœbe, Barrosa and Racoon and formed part of the expedition.

The medal is the same pattern as that issued for the Ashantee War but without any clasp, the name of the expedition being placed on the rim, and those officers and men already in possession of the medal for former operations will be entitled to have 'Mwele 1895' engraved thereon.

I have been unable to trace the authority for the engraving of 'Mwele1895–96' but perhaps the above notice was intended to cover 1896 as well.

This campaign was against the Arab Mazrui tribe who lived along the coast and in the hinterland near Mombasa. When Mombasa capitulated to the Muscat Arabs in 1837 the tribe split into two. The Beni Othmans went south to Gasi and the Beni Zaher to Taka Langu in the north.

The coastal Arabs' prosperity was based on their owning large quantities of slaves who worked their plantations of maize, millet, cassava, mangoes, oranges, coconuts etc. When therefore I.B.E.A.Co. decreed that all slaves should be freed, it dealt a severe blow to the Arabs' economy as the thought of hiring Africans on a wage basis was beyond their comprehension. This led to a decline in their prosperity and exasperated their feelings towards the British who had caused their downfall.

When the then Sheik of Taka Langu died in 1895, a dispute arose as to who should be recognised by the I.B.E.A.Co. Raschid was chosen and the unsuccessful candidate Mbaruk retired a few miles inland to Konjora where he seized the late Sheik's armoury and gathered together an army of 1,500 men. All this occurred at the time the I.B.E.A.Co. was handing over power to Her Majesty's Government, and the British Authorities under-estimated the threat involved.

A warship was sent to Kilifi and Sir Arthur Hardinge tried to negotiate with Mbaruk who refused to meet him. A force then marched inland on Konjora. Here Mbaruk's men fired on the Government troops which consisted of 310 seamen, 50 marines, 54 Sudanese and 160 Zanzibari regulars. War was then inevitable.

The Mwele campaigns which followed covered an area just over a hundred miles along the sea coast by 25 to 30 miles inland, rising to the steep hills 1,500 feet high in the south. Much of this area was covered with dense forest and intersected with creeks and swamps. Ideal guerrilla country but uncomfortable for the troops.

Mbaruk first moved north to the Sokoki Forest where he was joined by his brother Aziz. The Sokoki Forest must have been a good hide-out as it is still today dense thick forest, the home of elephant and buffalo, and in those days must have been impenetrable.

Aziz then attacked Taka Langu but was repelled.

An expeditionary force under the command of Admiral Rawson from H.M. ships St. George (Capt. Egerton), Phœbe (Capt. MacGill), Racoon (Comdr. Underwood), Barrosa (Capt. Marx) and Blonde (Comdr. Festing) consisting of 400 men (220 seamen and 84 marines), with 2 maxims and 7-pdr. gun, a rocket tube, 60 Sudanese, 50 Askaris and 800 porters, started inland.

Gasi, which is on the coast, was found to be deserted and in the meantime the rebels had attacked and burnt Vanga on the Anglo German border.

After a skirmish at Nololo, Mwele was reached on the 17th August.

Mbaruk's Mwele stronghold was located in the Shimba Hills which rise to about 1,500 feet, about 10 miles from the coast. These are now a reserve for the Sable antelope, probably one of the most spectacular beasts in Kenya.

Even today's visitor, driving often on a tarmac road in a comfortable car with an ice box of drinks, the climate is hot and even mild exercise such as catching the butterflies which exist in the forests in profusion, is tiring. Marching in thick clothes, carrying heavy equipment in pursuit of 'will o' the wisp' terrorist, must have been no picnic at all.

Mwele in 1895 was a strongly stockaded place and was taken after an engagement lasting two hours. The British casualties were 3 killed and 11 wounded. The only trace I could find of Mwele nowadays is a sign beside the road.

The Mazrui then commenced a series of guerrilla raids and for a period of four months or so held the initiative. In fact British control existed only in the ports. Malindi, now a tourist resort invaded by package tourists, was looted and the Indian shops burnt.

Reinforcements consisting of a contingent of 299 Indians for the E.A.R. arrived in December 1895, and in March 1896, the 24 Bombay Infantry (Baluchis), 749 all ranks, joined in the hunt.

A small naval contingent from H.M.'s ships Widgeon and Thrush also operated south of Mombasa. Their policy was to establish strong-points and deny the rebels any supplies.

The troops spent their time trying to catch the elusive Mbaruk and the campaign only ended when he and his followers fled into G.E.A., where on the 20th April 1896, 1,100 Arabs (of whom 600 were armed) surrendered to Major von Wissmann of the German Army.

Thus ended the Arab domination of the Kenya Coast and henceforth this area was firmly under British control.

P.R.O. file W.O. 100/90 contains a note 'those at Mombasa not to get medal.'

Entitlements:—

Uganda Rifles	329
Bombay Rifles	500
Followers	69

The Mwele medal is fairly common and as it is without a clasp, unexciting, but whilst the campaign it commemorates had no large pitched battles, it must have been extremely exacting for the troops.

CHAPTER SIXTEEN

Uganda

HISTORICAL BACKGROUND

'The Kingdom of Uganda, includes the provinces of Bupu, Singo, Ankole & Usoga. The Kingdom of Unyoro is about the size of Ireland & of Toro a quarter of its size.' (Thurston)

The Protectorate includes all three Kingdoms which we really do rule and the undefined regions of Nandi & Kamondo.

The coastal regions of the East African Protectorate (Kenya) have been referred to under Coastal Areas in the expeditions against the Arabs.

The inland portion of the Protectorate developed along the English Road to Uganda, a caravan route used by porters and donkeys from Mombasa via Ndi, Machakos, Fort Smith at Kikuyu, Ravine and Mumias to Kampala.

In 1890 the cost of portering a ton of goods from the coast to the Lake was between £150 and £300 a ton.

A start on the Uganda Railway was made in 1892 from Mombasa, but by 1898, at the time of the Uganda mutinies, the railway had only reached Ndi, 150 miles from Mombasa, and from there everything had to be head-loaded or carried by donkeys for the remaining 600 miles.

It was in 1892 that the railway was held up at Tsavo by man-eating lions, but the railhead reached Nairobi by 1900 and by 1901 Lake Victoria at Port Florence (Kisumu).

In 1898, a British Protectorate took over the I.B.E.A. Co, in 1902 Kenya became a Crown Colony and in 1963 became independent.

At the sources of the Nile on Lake Victoria in the middle of Africa existed the Kingdom of the Buganda. Little is known of their early history but they were a Hematic Race, probably originating from Abyssinia, called the Bahima. The tribe split into the Kingdoms of Rwanda, Burundi, Ankole, Toro, Bunyoro and Buganda.

The Buganda had a complex political system at the head of which was the Kabaka, an elected monarch, with a Queen Sister or Lubaga who was similarly elected. The parliament or Lukiko was headed by a Prime Minister or Katikiro and consisted of Nobles and the Governors of Provinces. The Army was commanded by a Field Marshal or Mujasi,

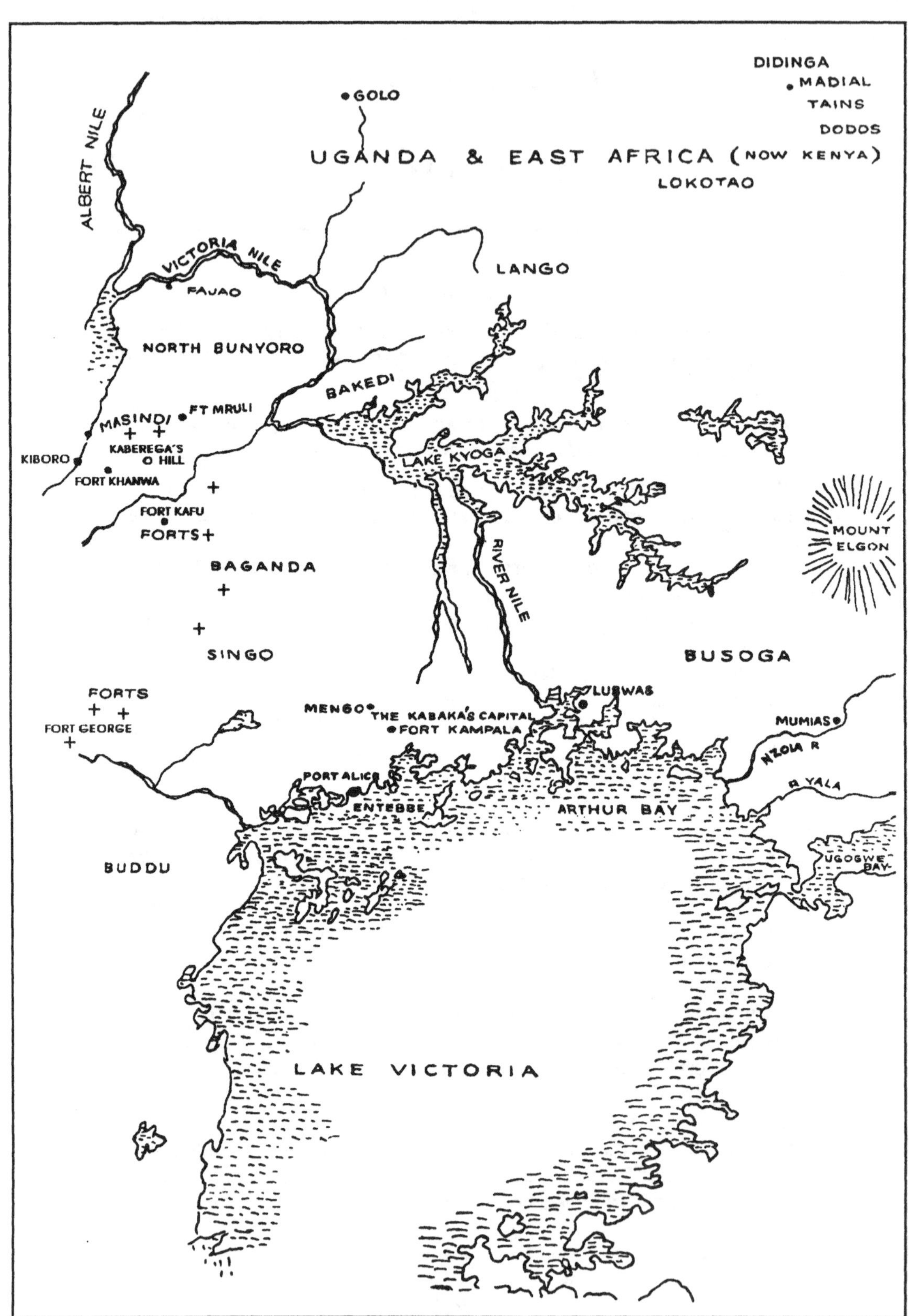

UGANDA & EAST AFRICA (NOW KENYA)
DIDINGA
MADIAL
TAINS
DODOS
LOKOTAO
GOLO
ALBERT NILE
VICTORIA NILE
FAJAO
LANGO
NORTH BUNYORO
BAKEDI
FT MRULI
MASINDI
KABEREGA'S O HILL
KIBORO
FORT KHANWA
LAKE KYOGA
FORT KAFU
FORTS
RIVER NILE
MOUNT ELGON
BAGANDA
SINGO
BUSOGA
FORTS
FORT GEORGE
MENGO
THE KABAKA'S CAPITAL
FORT KAMPALA
LUBWAS
MUMIAS
NZOIA R
PORT ALICE
ENTEBBE
ARTHUR BAY
R YALA
BUDDU
UGOGWE BAY
LAKE VICTORIA

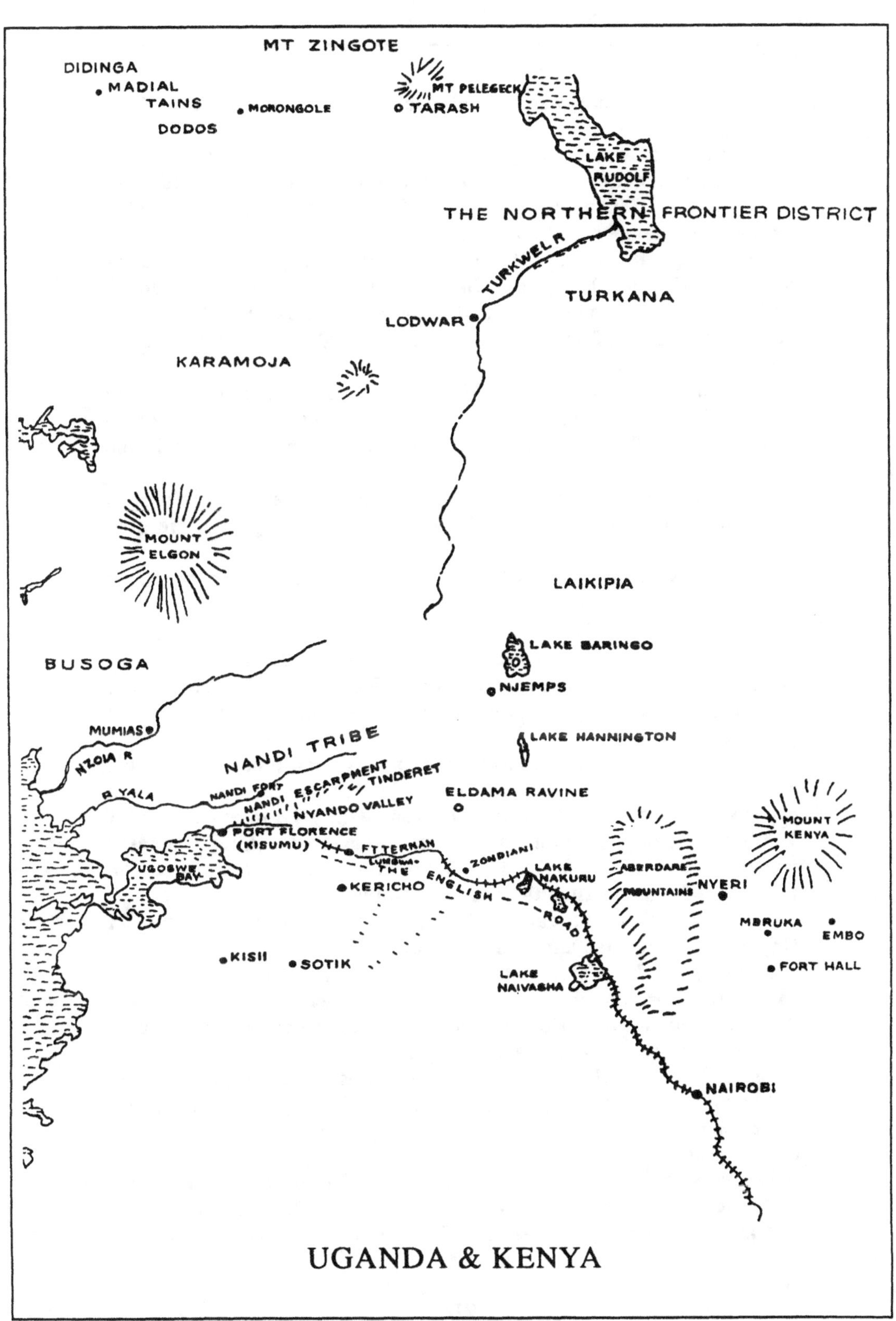

UGANDA & KENYA

whilst the admiral commanding the fleet of war canoes was called the Gabunga.

Their religion was a cruel pagan cult involving much human sacrifice and mutilation called Lubare. Not surprisingly there was a Lord High Executioner or Mukajanga who was a highly influential if somewhat overworked member of the Court.

At the beginning of the 19th Century King Kamanya gained victories over the tribes north of Lake Choga. In 1836 he was succeeded by Suna who was even more cruel and despotic than was normal and when he overran Ankole, Bunyoro and Usoga he slaughtered thousands after they had been disarmed.

In 1856 he was succeeded by Mutesa and was in this Kabaka's reign that outside contact was really made with the Kingdom. The first to penetrate were the Arab slaves who established trading depots in the southern portion of the Buganda.

The Zanzibar slaver Snay Bin Amir in fact arrived at the Court in 1852 and introduced Islam and sodomy to the Buganda. He also converted Mutesa to Islam.

The British explorers Speke and Grant visited Uganda and were received by Mutesa in 1860.

In the meantime, Egypt had been expanding since 1820 southwards down the Nile, undertaking the conquest of the Soudan to obtain ivory and slaves, and their rule eventually extended as far as Fashoda, where the Sudd of floating papyrus swamps blocked the further southward passage of the Nile.

In 1869 the Egyptian Government commissioned Baker to bring the area South of the Sudd swamps under Egyptian influence. Baker set off from Khartoum in the following year with a large expedition of 1,000 men, equipped with artillery and steamboats. After 4 months travel the expedition arrived at Gondokoro and Baker hoisted the Khedive's flag and called the province 'Equatoria.'

By 1872, Baker had reached King Kaberega's capital, Masindi, in Unyoro which he also annexed to Egypt.

Kaberega, who was a not mean fighter, objected, and sent his adversaries a present of poisoned banana beer, then launching an attack the following morning. He did not overrun Baker's camp but the latter was forced to retreat and abandon Unyoro.

In 1873 General Gordon succeeded Baker as Governor of Equatoria and using the steamers Khedive (108 tons) and Nyanza (38 tons) opened up communications from Duffile on the Upper Nile to Lake Albert and the Wamjoro.

It is remarkable to think that at one stage Gordon tried to launch a seaborne Egyptian expedition to Uganda from the Coast of East Africa through the Sultan of Zanzibar's territory, but on this occasion H.M.G.

sent a naval ship and forced the Egyptians to withdraw from Formosa Bay near Lamu.

General Gordon had negotiations both with Kaberega and Mutesa and when in 1878 he retired, he was succeeded by a German doctor, Eduard Schitzer, who had embraced Islam, tried to pass himself off as an Arab, and was known as Emin Effendi Pasha.

In 1878 Mackay of the C.M.S. arrived at Mengo and tried to convert Mutesa to Protestantism; he was followed in 1879 by the Roman Catholic White Fathers who in turn tried to persuade the Kabaka that theirs was the only true faith. The Muslims who had been the first to convert Mutesa of course condemned both types of Christianity. He, not unnaturally wishing to retain his independence, skilfully played off one faction against the other and probably remained a pagan Lubarist when he died in 1884.

Mwanga, a youth of 17, was elected Kabaka. He was a sodomite, cruel and weak, who indulged in the persecution of the Christian converts whom he had cut up and burnt.

Traditionally, it had always been believed that an enemy attack on Uganda would be launched from the east, and when in 1883 Bishop Hannington, who had come out to take over his diocese was trekking to Kampala, he was murdered on Mwanga's order at Lubwas. Perhaps Mwanga really believed it was the Church Militant coming to take over his kingdom. This naturally created quite a stir in Church and political circles in the U.K.

In 1884 came the rebellion by the Mahdi and the Sudanese against the Khedive, and Equatoria was cut off from the north. Emin Pasha had to retreat southwards with his troops.

As the Mahdi's forces moved ever southwards to extirpate the Turks, Emin's Sudanese troops mutinied and deposed him. In the meantime, Stanley led an expedition from Europe to rescue Emin, choosing the quite unnecessarily difficult route via the Congo from the West Coast, but eventually the remnants of Stanley's caravan arrived to escort a reluctant Emin to Bagamoyo in G.E.A., where they all arrived in 1889.

In 1888 Mwanga was deposed but was restored with the help of the Protestants in 1889.

In 1890 Captain Lugard, last heard of in British Central Africa, arrived with a few soldiers and much courage to impose I.B.E.A.Co.'s rule on the Kabaka. He made Mwanga, the Kabaka, fix his mark to a treaty and then carried out an expedition westwards, annexing Ankole and Toro where he joined up with Emin Bey at Kavallis to the west of Lake Albert.

After a civil war between the Protestants (Wa Ingresi), the Catholics (Wa Francosi) and the 'Wa Futi Bangi,' literally translated 'the marijuana smokers,' Lugard divided the country into three provinces apportioned between the three religious parties.

Lugard then returned to England, prior to his appearance in our story in West Africa.

In 1895 a Protectorate was declared over Buganda, to which Unyoro and Busoga were added in 1897, Ankole in 1899 and Toro in 1901. The north-eastern portion was annexed in 1902 as the Rudolf Province.

In 1903 a considerable portion of the eastern part of the Uganda Protectorate was passed to British East Africa and later the whole of Turkana was similarly transferred to present day Kenya.

In 1962 Uganda became independent.

The Foreign Office in view of the complaints made by the French Missions appointed Col. J. R. MacDonald to carry out an enquiry into Lugard's activities. Sir Gerald Portal was sent in 1893 to report on the desirability of abandoning Uganda and on 1.4.93 raised the Union Jack at Kampala and proclaimed Her Majesty's Protectorate over the Kingdom. Portal left Kampala on 30.5.93 after the death of his brother and proceeded to the west via the Tana River accompanied by Rhodes, Villiers, Berkerley & Arthur. Col. MacDonald appointed Acting Commissioner. Col. Sir H. Colvile arrived in Nov. 1893, with 2 Arabic speaking officers Capts. Gibbs & Thurston, but was later evacuated sick to the coast.

The casus belli was Kaberega had attacked our posts, cut off a party of our soldiers and raided Uganda. War was declared on 4.12.93 & Mwanga provided 15,000–20,000 men.

CAMPAIGNS IN UGANDA

THE CENTRAL AFRICA MEDAL WITHOUT CLASP

Army Order No. 66 of 1895 authorised the medal for the forces employed in Eastern Africa 'Unyoro December 1893—February 1894, Mruli April—June 1894.'
Army Order No. 100 of 1896 approved the medal for operations in East and Central Africa being granted to the forces employed during April and May in the Unyoro expedition.

In November 1893, Colonel H. Colvile (Grenadier Guards), the new British representative, arrived in Kampala, the Buganda capital, but later moved his headquarters to Port Alice (Entebbe) on Lake Victoria, which is at an altitude of 3,000 feet.

These campaigns were against King Kabarega of the Bunyoro, whose army was rumoured to consist of 8,000 men equipped with firearms and 20,000 spearmen.

The book 'Roddy Owen' sets the following picture:—

'Kaberega, King of Unyoro, was at least partially kept within bounds by the presence of the forts. This man scion of a long line of kings, had reigned for more than 20 years over the country commanding the trade routes between the Arabs on the north, and Uganda, the Congo and other districts on the South-west. He may have had knowledge that in days gone by Uganda had been the subject province to the Kings of Unyoro. Noted for his hostility to Europeans he had flouted in turn Bakero Casati attacked Stanley and during his long reign reduced to weakness almost all the surrounding tribes by raiding and murder. In his hands he held the salt deposits of Kibiro on the Albert Lake and the trade routes to the Soudan and by exchanging salt for arms and powder and the arms and powder for ivory and slaves had amassed great wealth and was able at any time to send several armies into the field. His subjects feared his sacred sceptre and bewitched spear. They worshipped him in good omen and would gladly die which they often did by hundreds if they could but look upon his face.'

Lugard describes Kabarega:—

'For years he had exhibited a continued hostility to all Europeans ... Later he outraged and insulted Major Casati when a guest at his court and bound him naked to a tree ... He fought against Stanley ... He gratuitously assisted the Mohammedan party ... He fought continuously against me ... He has sent an army against my Sudanese forts. I knew his overtures to be insincere.'

Lugard believed the Kabaraga was about to attack the forts in the Singo area. Major E. R. Owen (Lan. Fus.) with 200 Sudanese attacked Kaberega before he could get started and after a three hour fire fight, the Bunyoro broke and fled.

Kaberega however continued raiding into Toro and Colonel Colvile therefore set out to Bunyoro with a force consisting of 421 Sudanese with 156 armed and 200 unarmed porters and a 45 calibre maxim gun. He also had a collapsible steel boat, which was built in sections, for crossing rivers. The Buganda provided levies of 15,000 spearmen.

Against this overwhelming force Kaberega retreated northwards, but at Lendui he ambushed the advance guard and then retreated again into the Budonga Forest, an area of thick thorn bush, grass and ravines, with little or no supplies and where it was impossible to follow. Colvile's force built forts, hoping to blockade Kaberega, who eventually emerged from his fortress and was surprised by the Buganda when 300 of his stock and 40 guns were captured. The Bunyoro then fled across the Nile.

Kaberega next appeared in Eastern Bunyoro at Mruli.

Colonel Colvile with Captain J. H. S. Gibb (Worcs. Regt.) and a force of 200 Sudanese and over 5,000 Buganda levies again set out to capture or kill him. However Kaberega vanished across the Nile and Gibb had to return to Kampala.

Campaigning in the papyrus swamps of the Upper Nile was extremely arduous as papyrus floats and it is only too easy to lose one's footing and sink into the liquid mud between the clumps. With mosquitos galore and leeches in profusion, it cannot have been comfortable. Apart from these insect pests, crocodiles in those days were a problem and bit or carried off numbers of people, soldiers included.

Captain A. B. Thurston (Oxford Light Infantry) had a skirmish with Kaberega's forces in May 1894 but the latter then returned to Bunyoro and sent his army to threaten Hoima. With 150 Sudanese Thurston fought a pitched battle and defeated the Bunyoro who suffered 200 casualties.

Later in November Thurston almost caught Kaberega at Machudi, which he entered after dark, but Kaberega fled, leaving all his possessions behind.

In March 1895 the efforts to defeat Kaberega continued, and a large column from Hoima met a serious defeat, in spite of covering fire from a maxim, when they were trying to cross the Nile in native canoes. They were driven off by the heavy fire from the rebels' rifles. One British officer was killed and another wounded, apart from casualties amongst the rank and file.

A stronger expedition under Major G. Cunningham (Derbyshire Regiment) and Captain Ternan, consisting of three officers, 6 coys of Sudanese (500 men), 20,000 Waganda, 2 Hotchkiss guns and 3 maxim guns forced the Nile and defeated Kaberega, who as usual escaped.

Col. Colvile—28 Dec 1893 'The Army as a whole marched off together for the first time.' It consisted of:—

Europeans	8
Native Officers	14
Sudanese Officers & men	415
Walanda Great Chiefs	15
Walanda Riflemen	207
Walanda Musketeers	3308
Walanda Spearmen	10,600
Masai Spearmen	11
Swahili Headmen	8
Swahili Askaris	25

Swahili Porters (arrived)	209
Total Combatants	**14,820**
Swahili Porters (unarmed)	56
Enlisted Lendus	97
Sudanese followers	100
Walanda followers	1000
Walanda attached HQ	20
Wanyon Chiefs	2
Wanyon men (bearer company)	110
Wanyon interpreters & guides	13
Servants & interpreters	34
Artisans	3
Total non-combs.	1,425
Grand Total	**16,255**

The names of the Waganda Chiefs were engraved on their medals. P.O. despatch 24.4.95 to Acting Commissioner of Uganda and the unnamed medals distributed by Lt. Col. Ternan at Hoima, Kitwani & Kiboro.

Mr. Purkiss an ex-officer of I.B.E.A.Co. had been in the Merchant Navy and was in charge of the steel boat comprising 11 sections.

A Protectorate was formed in 1894, and the Civil Administrator's policy was to make friends with Kaberega. The final reckoning with this King was therefore deferred until 1899.

This is one of the few occasions on which a medal was awarded for a campaign when few, if any regular troops were engaged.

Roll W.O. 100/76 indicates that approximately the undernoted were entitled to the clasp:—

Principal Waganda Chiefs	21
Mr. G. Wilson writing to the Marquess of Salisbury on March 1898 said very many of the Chiefs have the Unyoro Medal, which they value most highly and more than one chief wore it on his breast when killed in action at Usoga.	
Sudanese Troops later Uganda Rifles	447
Troops with apparent Sudanese names & ranks such as Ombrasha and Nafar	111
Native officers, N.C.O.s and men (no rolls)	170/200

However, a note to the Roll states that 706 medals were supplied to the Foreign Office for distribution to the forces employed under Col. Sir H. E. Colvile.

Many of these medals must have been lost or destroyed in the 1897 mutinies and it is significant that Gordon mentions no Uganda recipients.

Vandeleur writes in his book that the Commissioner, Mr. Berkeley, held a parade of Sudanese troops at Kampala on 11th October 1895 and presented the medals to the participants of the 1st Bunyoro expedition of 1893. So some of the medals for Colvile's troops obviously were distributed.

Seventy-seven of the Central Africa Medals issued for the Unyoro and Mruli expeditions 1894–95 were returned to the Foreign Office and were sent to Woolwich for disposal in August 1899.

EAST & CENTRAL AFRICA MEDAL 1897–1899

Army Order No. 29 of 1899:—
Her Majesty has been graciously pleased to approve of a medal called the East and Central Africa 1897–1899 Medal being granted to the forces employed in the Military operations in the Uganda Protectorate in 1897–98.
The clasp 'Lubwas' will be granted to all Her Majesty's forces and allies who took part in the operations against the Sudanese mutineers from 23rd September 1897 to 24th February 1898.
The clasp 'Uganda 1897–98' to those who took part in the operations in Uganda other than those against the Sudanese mutineers from 20th July 1897 to 19th March 1898 inclusive or who reached Uganda within those dates.

Lubwas was a fort so named after the local chieftain, then spelt 'Lubas.'

The trouble started when Mwanga, the Kabaka of Buganda, who had recently been fined for smuggling guns, rebelled and went to Buddu where he was joined by many dissatisfied factions.

The Budda Field Force was assembled under command of Lt. Col. Ternan and comprised most of the available troops. After long marches he attacked Mwanga at a hill called Kabwoko, where after a hour's fight, Mwanga was defeated and fled to German territory at Bukoba.

Lt. Col. Ternan then placed Mwanga's infant son on the throne of Buganda. The troops in the Field Force received a month's extra pay and Queen Victoria's congratulations on the occasion of her Jubilee in 1897.

All then seemed well, but there was considerable dissatisfaction amongst the Sudanese who were then considered the best material for soldiery in Africa. Their pay at only 7/- per month was less than that earned by a caravan porter and was generally up to six months in arrears. Moreover, the Sudanese owed their allegiance to the Khedive of Egypt, and the British operations in the Sudan against the Mahdi had an unsettling effect. Unfortunately many of their British officers could not speak their language.

The immediate spark that set off the mutiny was the mounting of an exploratory expedition under Major (Lt. Col.) J. R. L. Macdonald to obtain more accurate information of the country north of the Uganda Protectorate but primarily to see if the French and Belgians were encroaching on the Nile.

The expedition was to concentrate at Njemps just south of Lake Baringo (in present day Kenya). This consisted of 6 officers and a detachment from the 14th and 15th Sikhs which was sent from India to escort the 450 Swahili porters. This escort was considered to be insufficient and the 4, 7 and 9 Coys. Uganda Rifles were also detailed to join the expedition as escort.

At this time the troops available in British East Africa under Major Hatch at Mombasa were:—

Indian contingent Punjabi Mussalmans	289
Sudanese	256
Swahilis	368
Arabs	16
Somalis and Goshas	191
	1120

Although the Sudanese soldiers had seen much service, their discipline was bad and dissatisfaction was rife, particularly amongst the Sudanese officers and N.C.O.s., many of whom had been demoted in rank for disciplinary reasons.

In September, when concentrating at Eldama Ravine, some of the men of No. 4 Coy. and all of No. 9 Coy. deserted. No. 7 Company, who had already reached Njemps, then joined the mutineers who concentrated at Eldama Ravine, where Macdonald tried to parley with them, without success.

The mutineers then marched westwards to Mumias and by-passed the fort, looting stations in the country as they went. From thence they went to Lubwas Fort on Lake Victoria. At Lubwas they hoped to join forces with the garrison and the dissatisfied Muslims of the Buganda tribe.

Major A. B. Thurston and N. Wilson arrived at Lubwas by steam launch before the mutineers, but they and Scott were seized by the garrison of the Fort, which was shortly afterwards joined by the mutineers from Eldama, who now numbered approximately 600 Sudanese and 200 Buganda Muslims.

On the 18th October Macdonald arrived at Lubwas with 10 Europeans, 17 Sikhs and 340 half-trained Swahilis. After an abortive parley, the mutineers attacked him for five hours, but were eventually driven off. Macdonald had 46 men killed and wounded in the action. Considering the untrained state of his Swahilis, and the fact that the mutineers were led by experienced native officers, this was a remarkable feat.

The mutineers then shot the three captive officers in cold blood.

The Foreign Office having been informed that the local forces were quite insufficient to deal with the situation, Capt. W. C. Barrett (1 Sikh Infantry) and 300 Sikhs of the East African Indian contingent were sent from Mombasa, and the 27 Bombay Infantry under Lt. Col. W. A. Broome, consisting of 12 officers and 737 sepoys were sent from India. These were later followed by a wing of the 4th Bombay Infantry.

All this took time and Col. Broome did not arrive in Kampala until July 1898 as the troops had to march most of the way and stores had to be carried by porter. There were few if any supplies on the 'English Road' between Mombasa and Kampala.

The mutineers were invested in Lubwas Fort by Capt. E. M. Woodward (Leic. Regt.) with a very small force, and after several skirmishes they evacuated Lubwas and moved by launch and canoes up the Nile towards Mruli.

In the meantime many of the Uganda Rifles' Sudanese troops had been disarmed and disbanded without causing trouble.

Reinforcements had arrived, consisting of: Major C. H. U. Price, commanding troops in Unyoro, 4 officers, 237 Indian ranks, 27 Bombay Infantry; 1 officer and 24 Askaris Uganda Rifles; Major C. Martyr, D.S.O., commanding 2 officers and 159 O.R.s Uganda Rifles; 1 officer and 53 askaris of the E.A.R.; 1 officer and 7 men of the 27th Bombay Infantry; and one 7–pdr. gun and three maxim guns.

The forces available were assisted by the Buganda, and the mutineers were chased around the countryside with actions at Kijembo, Kabagambi, Masindi, at all of which battles many of the Sudanese were killed. Operations were then continued against those chiefs in Bunyoro who had supported the rebellion.

Macdonald reported that he had fought five major and seven minor engagements and thirty-five skirmishes. Some 2,000 troops, 3,000 Buganda auxiliaries armed with firearms and many thousands of porters had been employed.

Casualties were 7 European officers and 280 African and Indian troops killed, and 5 Europeans and 555 African and Indian troops wounded.

The contents of Medal Roll W.O. 100/90 are confusing and very careful research would be needed to obtain really accurate assessment of how many are actually entitled to the medal and clasp.

On some rolls many are shown as entitled to clasp 'Lubwas' and not 'Uganda 1896/97,' and although this is corrected on some rolls, there are obvious mistakes on others. Careful cross checking for duplication is also necessary.

	1896/97	Lubwas
The Uganda Administration. Of these, 19 were awarded to the clergy (13 C. of E. and 6 R.C.) 7 ladies also received the clasp.	82	21
Officers and N.C.O.s, Misc. and a civilian volunteer (Mr. S. Ormsby, Merchant, Kampala)	99	–
Major Macdonald's contingent 14th and 15th Sikhs	49	48
27 Bombay Light Infantry	339	–
17 Baluch Light Infantry	536	88
Officers, N.C.O.s, and O.R.s Indian contingent	232	225
Uganda Rifles, with a note that only Europeans' and Indian's medals to be engraved	1074	55
5 Uganda Bn. K.A.R. Sudanese	117	–
Sudanese Company E.A. Rifles	150	150
Lt. Osborne's section of Swahili porters	29	29
Lt. Tracey's section	89	–
Surgeon Capt. McLoughlin's Swahilis	32	32
Lt. Fowler's Swahilis	59	59
Charles Bagnall's section—Swahilis	50	50
Capt. Ferguson's Swahilis	63	63
Lt. Malcolm's section	32	32
Capt. Austin's Swahilis	43	43
Capt. Kirkpatrick's section	55	15
Askaris and porters of Steamer Transport	60	60
Buddu Garrison	60	60
No. 2 Derajat & 4 Hazara Mtn. Bty.	5	–
Details of Indian Troops from varying regiments: 1 Sikhs, 2 Sikhs, 1 Punjab Infantry, 29 Bombay Infantry, 30 Punjab Infantry, 24 Punjab Infantry, 4 Punjab Infantry	59	19
Capt. Meldon's section	32	32
Chiefs and others of Buddu, Koki and Ankole	56	–
Similar as per Col. Macdonald's list	60	–
Waganda 'who did excellent service'	132	–

UGANDA 1899

(21ST MARCH—2ND MAY 1899)

Army Order No. 254 of 1900 approved the clasp inscribed 'Uganda 1899' being granted to Imperial and Colonial Forces who were employed in the military operations against Kaberega in the Uganda Protectorate between 21st March and 2nd May 1899.

This was the third and final round with King Kaberega.

Major G. C. Martyr (D.C.L.I.) advanced down the White Nile, establishing posts at Fajao, Affuddu and Lamogi.

Reinforcements of 500 under Lieut. C. U. Price arrived in December 1898 and Martyr was informed that Mwanga, with a force of 300 rebels and 50 mutineers from Lubwas, was at Kangia.

They crossed Lake Kwania by canoe and captured Mwanga and Kaberega. This action was described as 'an unexpected and unspectacular conclusion to two long and troublesome careers.'

In one account it is stated that Kaberega was wounded in his last fight and taken to a military hospital, but as some other African was treated before him, he became incensed, jumped off his stretcher and kicked the European doctor in the bottom. The doctor however took the assault philosophically,'After all, it's not every day that one is kicked by a king.'

Kaberega was exiled in the Seychelles but allowed to return to Uganda where he died in 1923.

A photograph of Mwanga and Kaberega taken at the time shows them as two rather inoffensive, middle aged men wearing kanzus, a sort of long cotton shirt similar to a night shirt.

Roll W.O. 100/90 indicates that the following were entitled to this clasp:—

Major Martyr and other Officers	13
Uganda Rifles:—	
Askaris	216
Indian ranks	190
Followers	15
	434

49 Temoseyo Mulondo ex Col. Macdonald's list—Lubwas clasp only. These medals were also issued in bronze.

African General Service Medal

LANGO 1901

Army Order No. 234 of 1902 authorised to Imperial and Colonial Forces employed in military operations in Uganda the medal with clasp 'Lango 1901,' to all officers and men who took part in the operations against the Lango people and Soudanese mutineers, under Major C. Delmé Radcliffe, Connaught Rangers, between the 24th April 1901 and the 24th August 1901.

Captain C. Delmé-Radcliffe (Connaught Rangers) assumed command of the Nile Military District with the role of pacifying the Lango District and hunting down the surviving mutineers. Like many working in Africa, Delmé-Radcliffe was given an African name. His was 'Langa Langa' or 'were-lion' as he made, to the Africans, incredible night marches with his patrols.

It was discovered that the hard core of resistance in his district was some 100 mutineers known as the 'Black Turks' who had entered into a blood brotherhood with some of the Lango who believed that King Kaberga would return to oust the British.

The Lango Field Force totalled 6 Europeans and 405 native officers, N.C.O.s and askaris from the 1, 3, 5 & 8 Coys. Uganda Rifles (4 K.A.R.) and a levy of 100 Buganda.

The country was thick forest and tall elephant grass up to nine feet high, and the campaign took place during the rainy season.

The mutineers had established a European model stronghold and were still well stocked with arms and ammunition, but their camp was captured at the commencement of the expedition.

Thereafter, the operations consisted of constant patrols in terrible conditions, having to rely on the country for supplies. During the campaign, all but seven of the mutineers were killed or captured.

The Lango tribesmen broke away from their blood brotherhood with the mutineers, for which purpose the medical officer of the column had a special de-oathing ceremony that involved injecting an emetic into the scar caused at the blood brotherhood ceremony. The resultant vomiting convinced the recipients that the evil spirit was departing.

The Field Force suffered 21 killed and 16 wounded and themselves captured 1,485 prisoners, 10,000 cattle, goats and sheep, 3,000 spears and 88 firearms.

W.O. 100/391 shows the following entitlements:—

Major C. Delmé-Radcliffe
Capt. C. I. R. Petrie
Capt. G. M. H. Harman

Capt. P. H. Walker
Capt. T. N. Howard
Capt. A. J. Whittle
Doctor A. G. Bagshawe Uganda Medical
4750 Serjt David Moon
Mr. (Capt.) Seymour Lecti (a collector) not entitled

4 Bn. K.A.R.	381
Buganda Levy	101
Total	490

This concludes the campaigns against the Sudanese mutineers.

CHAPTER SEVENTEEN

The Uganda Star

I am indebted to Mr. E. C. Joslin of Spinks for a copy of the Uganda Journal containing an article by E. F. Twining, later Lord Twining, the Governor of Tanganyika, on the Uganda Medals and decorations.

I quote what he wrote about the Uganda Star.

'By far the most interesting decoration that has been awarded to natives in Uganda is the "Mutiny Star." Few Europeans now in Uganda are aware of its existence, much less of its history and among natives its significance has been lost save to the remaining recipients or their immediate relatives. It is not surprising that the decoration is literally unknown to numismatists. The suggestion that a small silver star, to be worn like a medal, should be given to native chiefs emanated from Major Macdonald in May 1898 and was complementary to his proposal for a special war medal. It was to be awarded to those chiefs who had especially distinguished themselves during the military operations known as the Uganda Mutiny, not by actually taking part in the fighting but by rendering valuable service in assisting to check the Mutiny. The Commissioner, Mr. E. J. L. Berkeley, supported the suggestion on the grounds that the Uganda Chiefs had behaved throughout a critical time with so much courage and loyalty. Two lists were forwarded to the Foreign Office, one drawn up by Macdonald and a supplementary one by Mr. George Wilson, who had been acting Commissioner during the most critical months of the Mutiny. Macdonald's list included the names of 13 Baganda and two Basoga chiefs and 8 Sudanese officers, while that of Wilson contained the names of Mbogo, uncle of the King of Uganda, Kasagama, the King of Toro and Namswaga, the ruler of Koki, together with 9 Baganda and 3 Banyoro chiefs. The principle that the Star should be given for services other than in the actual fighting does not appear to have been strictly followed, for all but two of the Baganda on Macdonald's list were also 'mentioned' in Macdonald's despatches on the military operations and thus qualified for the Mutiny Medal, while two of the Sudanese were specially recommended for conduct during specific fights and one (Ferag Effendi) for saving the lives of Captain Molony and Mr. Malek at Jinja in October 1897. In February 1899 Her Majesty the Queen sanctioned the proposal and the Stars were made by Messrs.

Carrington and despatched to Uganda in time to be presented to the officers on parade and the chiefs in baraza on the Queen's birthday. Unfortunately the names of the six other Sudanese (one of whom was Yuz. Rehan Rashid, Capt. Sitwell's native officer in Toro) were somehow not included in the despatch that went home and it is not clear whether the error was ever rectified.'

The Star is of silver, eight-pointed, and in the form of a brooch. On the obverse is Queen Victoria's head (wearing a veil and diadem) surrounded by a scroll with the dates 1897 above and 1898 below the Queen's head. The name of the recipient is engraved on the back.

The full list of recipients is:—

MACDONALD'S LIST

Apolo, Katikiro	Kyramia (Msoga)
Mugwanya, Katikiro	Lubwa (Msoga)
Kakunguru	Jumba Gahunga (Roman Catholic)
Zakariah, Kangao	Mujasi (Protestant)
The Pokino (Alekisi Sebowa)	Mujasi (Roman Catholic)
The Sekibobo	Joshua Mugema
The Kago 'Yakobo Musajalumbwa)	Kisibika
Andrea Mukubira	

SUDANESE NATIVE OFFICERS

Hassan Effendi	Surur Effendi
Mahommed Ratib Effendi	Idrus Effendi
Juma Effendi	Ali Momboa Effenda
Abdalla Effendi	Ferag Effendi
Rehan Rashid Effendi	

WILSON'S LIST

Paulo Mukwenda	Mesisala Mukabya
Luwekula	Gideon Mtanda
Alenni Bugala	Ishmael Mlemna
Edward Mlinda Mzigo	Henry Wright Duta
Teofiro Musalosalo	

UNYORO CHIEFS

Abaswezi	Kiza
Rejumba	

I have not been able to trace any correspondence regarding the Star having received Royal approval. Nor have I met anyone who has ever seen this rare decoration. Nevertheless it certainly merits inclusion in the catalogues as a campaign decoration.

A photograph of the Native officers of the 4th K.A.R. taken in 1925 shows Bimbashi (Major) Ali Effendi Mombur D.C.M. one of the Star holders, seated next to the Duke of York (later H.M. King George VI).

CHAPTER EIGHTEEN

Campaigns Against the Nandi Tribe

There were four expeditions against the Nandi tribe, but only two medals were awarded, which is surprising as elsewhere medals and clasps were awarded for much less activity than that which took place against this tribe.

1st Expedition 1895
2nd Expedition 1897
3rd Expedition 1900 African General Service Medal, with bar 'Uganda 1900' (3rd July—October 1900)
4th Expedition 1903
5th Expedition bar 'Nandi 1905–06' 18th October 1905–6th July 1906)

The Nandi tribe formerly ranged over a great area and in 1890 Lugard wrote, 'to the north would lie the country of Nandi, whose people are reported to be excessively hostile and fierce, and no caravan, European or native, has ever yet dared to cross their hills.'

The Nandi were the strongest of the Kalenjin group of tribes which includes the Kisii, the Kipsigis and the Elgeyo. They are related to the Masai.

The overall spiritual head was called the Laiban, but there was no Nandi tribal chief and the tribe was divided into porosieks, which consist of fighting groups as well as being political units. The tribe was pastoral and were armed with bows and arrows, spears and the usual shields of oxhide. They practised circumcision and worshipped the sun and when a man died, his wives were buried with him.

Their country overlooks the Nyando Valley with a massive, almost cliff-like escarpment rising to approximately 6,000 feet, from the plain, where the old caravans to Uganda marched, and later the Uganda Railway ran. Their country is heavily forested and much was recently still virtually undeveloped, as quite a large area had been allotted by the Government to Lord Kitchener of Khartoum at the beginning of the century and had remained in its virgin state until the 1950s.

The Nandi, being a belligerent tribe, had boundaries ranging over a large area to include the Kavirondo Gulf to the Masai land near Lake Naivasha, which at this time was part of the Uganda Protectorate. Now all the Nandi tribal areas are in Kenya.

The caravans, which consisted of parties of porters, for everything had to be carried by humans on their heads or on donkeys passed within sight of the Nandi Escarpment and such easy pickings were a considerable temptation to the Nandi, to which they often succumbed. A typical foray was against a Scots trader called West who had pitched his camp near the Escarpment. The Nandi attacked and killed him an 23 of his porters and looted his camp.

1ST EXPEDITION

The mail also came under attack and in October 1895 Major Cunningham of the Uganda Rifles led a force of 1,000 against the Nandi.

There was some skirmishing when the Nandi attacked with spears, for although they had captured rifles, they could not use them. Vandeleur describes a charge made by 500 WaNandi right to within 30 yards of the Sudanese troops. Many of the tribe were killed but the British lost only 14 killed.

2ND EXPEDITION

'Peace' was established in February 1896 and a post set up at Nandi Fort about 4½ miles west of Kapsabet. By the end of the year, due to further raids and murders, a further expedition was required. This set out in 1897 and captured some 238 cattle and 7,838 sheep and goats. The casualties sustained by the Field Force were 6 askaris and porters killed and 5 wounded. 6,000 rounds of small arms ammunition were discharged as the troops were somewhat 'trigger happy.'

At this time Snider rifles had just replaced Remingtons and they in turn were being superseded by the issue of 450 Martini Henry carbine rifles.

3RD EXPEDITION

UGANDA 1900

(3RD JULY–OCTOBER 1900)

Army Order No. 133 of 1902 authorised for Imperial and Colonial Forces who were employed in the military operations in the Nandi Hills, the medal with clasp 'Uganda 1900' to all officers and men who took part in the actual operations in the Nandi country, between the 3rd July and October 1900 under Lieut.-Colonel J. T. Evatt, D.S.O., Indian Staff Corps.

The Nandi had remained hostile and the situation was aggravated by the approach of the Uganda Railway. Persistent looting and attacks necessitated another expedition. This was commanded by Lieut.-

Colonel J. T. Evatt, D.S.O., who was described as 'brawny, thick and glaring, a bit gray and a face like an old walrus, running them all with an iron hand ...'—he was almost a prototype of an indomitable 'C.O..'

In June, Captain Parkin (Northants. Regt.) and a small expedition of 25 rifles and some Masai spearmen killed 25 Nandi and captured 229 cattle and 1,800 sheep and goats. However they were followed up by the Nandi and lost 2 killed and 1 wounded.

The action precipitated Evatt's expedition which absorbed virtually all the resources of the Uganda Rifles.

The Nandi dispersed and refused to fight, but cut up a number of small parties, even annihilating 20 Sudanese in one skirmish.

Evatt was reinforced by Masai irregulars, who were adept at finding the cattle which the Nandi had hidden in the caves of the Escarpment. Thus quite severe pecuniary losses were inflicted on the Nandi as large numbers of cattle and goats were captured.

Evatt's camp was attacked and might well have been overwhelmed if the Nandi, who fought with great courage, had not been driven off by fire from the maxim gun.

At that time, Sir Clement Hill, the permanent Under Secretary of the Foreign Office, was at Fort Ternan, and as the railway was being delayed by the activities of the Nandi, a political solution was sought, the expedition recalled, and the Nandi bought off and appeased.

Casualties had been heavy on the Nandi side, 74 killed and 1,039 cattle and 3,100 sheep captured, while the Army's and auxiliaries' casualties had been 103 killed, 4 died of wounds, and 111 wounded. Sixty of the wounds were caused by arrows. Although the arrows were generally poisoned, none died of the toxic effects in spite of the fact that treatment consisted of injecting strychnine into the wound.

File No. W.O. 100/91 shows that this clasp was awarded to:—

British Officers	40
British N.C.O.s	15
Askaris	593
Indian Troops	225

These were from the 1st Bn. Uganda Rifles (4 K.A.R.) and the Uganda Rifles Maxim Gun Squads. There were also 20 Kisumu Police.

The total recipients appear in the region of 893 all ranks, far more than the 240 odd quoted by Gordon.

As was usual at that time, the Indian troops in the Uganda Rifles came from virtually every Sikh infantry unit in the Indian Army.

On the 1st April 1900 the Eastern Province of Uganda was transferred to the East Africa Protectorate (Kenya).

4TH EXPEDITION

The Nandi continued to murder personnel and steal Railway property, including a special bolt which was ideally suited to bash enemy's heads. The thefts reached such proportions that 100 miles of telegraph wire was stolen at one time.

In April 1903 Captain W. K. Nicholson (Indian Staff Corps) led an expedition of 7 officers and 580 regular troops from the 3rd, 4th and 5th K.A.R. against the Kamelilo section of the tribe. Eventually, after 300 head of cattle and 4,500 goats and sheep had been captured and 100 warriors killed, the tribe sued for peace.

As before however, the tribe committed further raids and murdered more European settlers.

5TH EXPEDITION

NANDI 1905–06

(18TH OCTOBER 1905–6TH JULY 1906)

Army Order No. 92 of 1909 amended the original order of 1906 authorising the clasp for the 5th Nandi Expedition as follows: 'The medal with clasp 'Nandi 1905–06' will be granted to all officers and men who served under the command of Lieut.-Colonel E. G. Harrison, D.S.O., Reserve of Officers; Captain (temporary Lieut.-Colonel) J. D. Mackay, Seaforth Highlanders (Ross-shire Buffs, the Duke of Albany's), and Brevet-Major (temporary Lieut.-Colonel) H. A. Walker, The Royal Fusiliers (City of London Regiment), respectively and took part in the operations in the Nandi country between the 18th October, 1905 and 6th July, 1906, both dates inclusive.

Lt. Meinertzhagen of the 3 K.A.R. was stationed at Nandi Fort. He made friends with the Nandi, who he obviously liked, as he described them as 'the most charming savages he'd met who should be taught a lesson.' They tried to ambush Meinertzhagen, as the Chief Witch Doctor, the Laibon called Koitalel, wanted his body to make a particularly potent medicine to turn bullets into water.

The Laibon Koitalel was believed to be the principle cause of the trouble.

Meinertzhagen expressed the view that the military wanted an expedition just so they could win a medal, and the politicals so they could absorb the Nandi land into the White Highlands, and he conceived the idea that if he could capture the Laibon then there would be no necessity for an expedition. Eventually he received orders to try to do this. He met the Laibon for a parley, but the Laibon tried to ambush him. The Laibon, together with 23 of his entourage were shot by Meinertzhagen's Askaris.

Opinion was the Meinertzhagen should be awarded a Victoria Cross, but even though there were three Courts of Inquiry which cleared him from acting improperly, the civilians had the last word as he was sent back to England. It was one of the many Administration versus the Army 'fetinas' (feuds).

This act, which occurred on the first day of the campaign, knocked the stuffing out of the Nandi, whose defences thereafter lacked coordination.

The expedition under command of Lt. Col. E. G. Harrison was divided into four columns: Col. Gorges from Londiani, Major Walker from Lumbwa, Capt. Maycock from Muhoroni and Capt. Barrett from Nandi Fort, and consisted of 540 men from the 1st K.A.R., 780 men from the 3rd K.A.R., 260 Armed Police, 1,000 Masai levies, 100 Somalis, 500 armed porters, 8,460 unarmed porters and 10 machine guns.

Against this force the Nandi had their traditional weapons and a few firearms which they still hardly knew how to use.

The campaign was nearly over in December 1905, but hostilities were renewed. The Nandi had 750 killed and 30,000 cattle and goats captured, whilst H.M.'s Forces' final casualties were 90 killed and wounded.

The Nandi were moved into a smaller reserve and thereafter no further trouble occurred.

File W.O. 100/390 shows the following as entitled:—

1 K.A.R., all ranks	598
3 K.A.R.	761
Indian contingent 4 K.A.R.	260
E.A. Police	422
Medical	12
Political	24
Miscellaneous, including a parson	8
Total	2,085

EAST AFRICA 1905

(31ST MAY–9TH OCTOBER)

The medal with clasp 'East Africa 1905' was granted to all officers and men under the command of Captain and Brevet-Major L. H. R. Pope-Hennessy, D.S.O., The Oxfordshire Light Infantry, who took part in the operations in and near Sotik, from the 31st May to the

12th July 1905, both dates inclusive, also to all officers and men under the command of Captain E. V. Jenkins, D.S.O., The Duke of Wellington (West Riding Regiment), composing of the Kisii patrol, which entered the Kisii country on the 1st September 1905 and patrolled it until the 9th October 1905, both dates inclusive.

This clasp was awarded for two expeditions to the Sotik and Kisii country.

As a safeguard to the Uganda Railway a military post was established in Kipsigis country at Kericho. This town, which is now the provincial capital and centre of the Kenya Tea Industry, is situated at 6,500 feet elevation in an area of particularly heavy rainfall.

The tribes to the south do not appear to have been particularly badly behaved, but in the spring of 1905 the people of Sotik raided the Masai, took prisoners and cattle, in fact, the sort of activity in which they had probably indulged from time immemorial. However, this time they refused to obey the Sirkari's (Government's) order to release the captives, and the Secretary of State gave permission for an expedition to be mounted against them and for their country to be brought under control.

The Kipsigis tribes are akin to the Masai and are a pastoral and Kalenjin tribe.

The expedition was commanded by Major L. R. H. Pope-Hennessy, D.S.O. (Ox. & Bucks Light Infantry) and was organised into two columns: No. 1 at Njoro 3 coys 3 K.A.R., a detachment of 1 K.A.R. and 600 Masai levies. No. 2 at Kericho, under Commander C. L. Barlow (West Yorkshire Regiment)—1 coy 3 K.A.R., 30 police and 300 Masai levies.

No. 1 column met with slight opposition and joined forces with No. 2 at the Sotik post after a week.

The usual punitive expedition procedure was followed and the force patrolled and eventually captured 5,000 head of cattle which they marched back to Molo, cutting a road 35 miles in length through the very heavy forest that exists in that area.

The casus belli, the Masai prisoners, were released.

The expedition against the Kisii was a more sanguinary affair. They had murdered some Kavirondo, a lake people quite unlike the Kalinjin pastoral tribesmen. In September, Captain E. V. Jenkins (West Riding Regiment) and a political officer set out from Kericho with 1 coy of 3 K.A.R. and some armed police into the Kisii country which adjoins Sotik.

The tribesmen, who were armed with long spears with small heads, did not even know what firearms were and thought that the Askaris were carrying sticks. They attacked right up to the K.A.R.'s bayonets and lost 67 dead.

A few days later a force of 600 Kisii again attacked and suffered another 70 or so casualties.

W. Lloyd Jones describes a later expedition which arose from the spearing of the District Commissioner. It was not a severe wound as he was out again in about a month.

The then orders were to:—

(a) Fire on all natives seen with weapons in their hands;

(b) Collect all cattle and goats.

1,000 huts were burnt and 150 Kisii casualties inflicted. He describes the 300 K.A.R. and 200 armed police as 'like a piano going up a narrow stairway when they tried to form square and march through the bush.'

File W.O. 100/394 shows the following as entitled:—

Provincial Commissioner C. W. Hobley
J. J. Drought, Intelligence
J. Corbet Ward
Lt. F. W. Maycock
Lt. Col. J. D. Mackay
Major L. H. R. Pope-Hennessy
Major E. V. Jenkins
Capt. F. A. Dickinson
C. L. Barlow
W. E. H. Barrett
J. K. T. Whisk
D. McLeod
Hon. L. P. Cary
Lt. J. H. Levonson Gower
Political Offr. G. A. S. Normcotel
J. W. T. McLellan
H. B. Bodeker
D. Stewart
E. W. Rodriod
Tpt. Offr. H. Rayner 3 K.A.R.
Sir Claude de Crespigny, Bt., 3 K.A.R.
Inspector J. H. Milton E.A.P.C.

All ranks 3 K.A.R.	336
E.A. Police	135
Total	493

CHAPTER NINETEEN

Expeditions against the Didinga and the Turkana in the Northern Frontier Districts

The following clasps are awarded:—
    East Africa 1913
    East Africa 1914
    East Africa 1915
    East Africa 1918
for these expeditions in the Northern Frontier Districts.

Army Order No. 353 approved for services in East Africa and Uganda Protectorate 1913–14 in the undermentioned military operations in the North Frontier Districts of British East Africa and Uganda Borders west of Lake Rudolph provided that the claims of officers, soldiers and others are approved by the Army Council, the medal with clasp 'East Africa 1913' will be granted to all officers, soldiers and others under the command of Captain W. T. Brooks, Duke of Cornwall's Light Infantry, who took part in the operations against the Didinga Tribe between the 17th June 1913 and 7th August 1913, both dates inclusive.

EAST AFRICA 1913

The Didinga were a large tribe of cattle raising and cattle stealing people inhabiting a range of mountains in the Sudan on the northern borders of the Uganda Protectorate. The Didinga were generally hostile to their neighbours and their country was virtually unknown for it was said that no trader, either native or European, had ever visited their mountains. the Soudan Government gave permission for an expedition to take place within their territory as the raids against the Dodo Tribes could no longer be tolerated.

An expedition under command of Captain W. T. Brooks (D.C.L.I.) was despatched to exact compensation for recent raids on the Dodos. The Didinga country was quite unexplored and unknown and the expedition left Madial without any previous reconnaissance or knowledge of their objective.

The force consisted of 190 rifles from the 4 K.A.R. divided into three columns:—

No. 1—Lieut. M. G. B. Copeman (Leic. Regt.) with 60 rifles and a maxim;

No. 2—Capt. R. H. Leeke (Rif. Bde.) with 60 rifles;

No. 3—Capt. W. T. Brooks with 70 rifles and a maxim.

The plan was for No. 1 column to act as a stop whilst the others were to beat out the country. Communication was by helio.

Initially surprise was achieved at the first village and some tribesmen were killed and stock captured.

On the second day of the expedition all three columns began driving the hills, over an area of about 24 square miles. As the land was intersected by deep gullies and the forest was heavy, the country was ideal for ambushes and defensive tactics and the Didinga soon overcame their initial surprise and fought back. Their tactics were to ambush from close quarters, and for arms each tribesman carried five to six spears. With these tactics by day, and with night attacks on the bomas containing the captured cattle, Capt. Brook's force must have had a roughish time, particularly when it came to extracting his force together with 2,000 cattle and 11 badly wounded men to Madia, and their withdrawal met with constant attacks.

The force was back at its base at Madial on the 7th August with 2,037 head of cattle, 1,660 goats and 62 donkeys. The force had lost 3 killed, and 10 men wounded, but the Didinga remained successfully subdued.

The roll shows 265 entitled to the clasp from the 4 K.A.R. battalion.

EAST AFRICA 1914

Army Order No. 353 of 1917 also authorised the medal with clasp 'East Africa 1914' to all officers, soldiers and others under the command of the late Captain R. H. Leeke, the Rifle Brigade, Lieutenant (local Captain) H. A. Lilley, the Yorkshire Regiment, who took part in the operations against the Turkana Tribe between the 2nd April 1914 and 7th July 1914, both dates inclusive.

This was the first of three expeditions against the Turkana Tribe, a nilo-hematic race living in the low lying areas in the north of Kenya, in arid, dry and hot conditions in a tribal area of about 20,000 square miles.

Turkana riflemen were described as:

'Very strange they looked, all stark naked except for red leather bandoliers' ... 'They had as have so many inexperienced natives a theory that to see an enemy, point their rifle towards him, press the trigger, should seal his doom. When this brave hope proved unfounded a distrust in their own firearms resulted.'

The fighting men, in addition to their spears and shields carried a a wrist knife, a circular band of iron of razor-like sharpness, a thin hollow scabbard of stout hide protected its owner from accidental injury.

The Northern Turkana—by far the most troublesome and dangerous of all the negroid races, their raids and atrocities. The principle commodity to be gained human beings, to be sold in slave markets cattle and ivory. The victims were generally the better behaved adjoining tribes, Karawoja, Suk, etc.

Regular Abyssinian troops uniformed and equipped were regularly crossing the frontiers into the Soudan, Uganda and came and went as they pleased. The Northern Turkana became their vassals.

By 1913, nearly the whole of the East Africa and Uganda Protectorates were administered, except for this area of the Northern Frontier District, (N.F.D.).

The N.F.D. was not easy country for campaigning as it was a foodless, waterless desert, largely unknown, very hot and in places covered with thick scrub.

The area west of Lake Rudolf was a 'Badlands,' a resort of Swahili ivory poachers, Abyssinian gun runners and lawless Turkana. Operations in the dry season were therefore limited by the availability of water to the areas around the Turkwel River and Lake Rudolf.

The K.A.R. had posts at Madial, Lokuta, Morongole and Logire and Captain R. H. Leeke (Rifle Brigade), having set up a depot at Kiteng, started to collect supplies and donkeys for transport for the forthcoming expedition.

It was reported that Lolel had acquired a number of Legras rifles and that another hostile chief, Ebe, had moved north to the area near Lake Rudolf, which the Turkana considered to be in Abyssinian Territory.

Captain Leeke's plan was to attack in two columns from Morongole and Lokuta on Mount Pelegech. However, in March, raids by the Turkana near Mount Zingote necessitated punitive action and Captain S. W. H. Silver (Suffolk Regiment) chased and caught the raiders, killing nineteen and recapturing some of the stolen stock.

In April an expedition set out from Tarash to capture the Turkana cattle said to be grazing on the River Tarash. Nothing was found there, but tracks led them to Mount Pelegech, an extinct volcano, where they saw some camels. The mountain was attacked and after a fierce fight with the Turkana when many were killed, 5,000 cattle, camels donkeys and sheep were captured, some of which turned out to belong to Lolel himself.

The column was off in pursuit of Lolel towards Lake Rudolf when orders were received from the Chief Secretary to stop operations because of the expedition against the Mareham in Jubaland and by the 2nd of May Leeke was back at Tarash.

Owing to the open nature of the land, casualties had been very low, consisting of only one porter killed and one askari wounded.

This retreat naturally encouraged the Turkana who although had a total of 8,000 stock captured, were unchastened and continued their raiding. In one such raid, some 500 Turkana killed 14 men and massacred 96 women and children in cold blood. They were brought to action by Capt. H. A. Lilley and 150 were killed in a brisk fire fight and much of their booty was recaptured. Another raiding party was ambushed in June when 57 raiders were killed.

Captain Leeke was about to set off on a major expedition when the war with Germany started and the K.A.R. were required for other roles.

The Roll shows that a total of 265 were eligible for this clasp.

Lolel described in 1918 as the Chief Witch Doctor was killed in fighting in 1918 at the Modo Pass.

EAST AFRICA 1915

Army Order No. 17 of 1917 authorised the clasp 'East Africa 1915' being granted for operations in East African and Uganda Protectorates to the forces employed in the military operations under Lieut Colonel W. F. S. Edward, D.S.O., against the Turkana tribes residing in the North Frontier Districts of British East Africa, and on the Uganda borders west of Lake Rudolf between the 4th February 1915 and the 28th May 1915, both dates inclusive, provided that the claims of officers, soldiers and others are approved by the Army Council.

To take over internal security from the Regular Army the Kenya Police Service Battalion was raised in 1914 with a strength of 300 Askaris. The Police were in action against the Southern Turkana at the foot of the Laikipia Plateau, where much stock was captured and taken to the Government pound at Kacheliba. This affair did not affect the Northern Turkana.

It is hard to understand why a bar was awarded for these punitive measures as the general lawlessness continued and there were several minor punitive expeditions during the following years, and in 1917 at Nakot Pass, Captain Rayne's square was attacked for four hours by 250 riflemen led by an Abyssinian with a trumpeter. Eventually the attack was beaten off and 37 dead raiders were found, but no medals were awarded for these activities.

The Medal Roll shows the following were entitled to this clasp:—

9th Sudanese	133
East African Police	198
Special Police	49
N.F.D. Constabulary	13
Sudanese Medical Corps	6
Uganda Police	62
Donkey Transport of E.A. Police	30
Total	491

The Roll for the Sudanese is signed by the Acting Sirdar of the Egyptian Army.

EAST AFRICA 1918

Army Order No. 51 of 1920 authorised the clasp 'East Africa 1918' to all officers and men who served under the command of Major R. F. White, Essex Regiment, Major H. Rayne, M.C., King's African Rifles and Captain J. H. R. Yardley, D.S.O., Royal Inniskilling Fusiliers, in the military operations against the Northern Turkana, Marille, Donyiro and kindred tribes, in the vicinity of the Southern Sudan boundary to the west of Lake Rudolf between the 20th April 1918 and 19th July 1918, provided that the claims are approved by the Army Council.

PATROL 44

This was a combined operation of the Soudan and British East African forces, the former operating from Mongalla—Torit.

Major H. Rayne officer in charge of the Turkana military district & the B.E.A. forces commanded the K.A.R. His temporary rank was relinquished when in the Soudan. Major R. F. White commanded the Soudanese forces.

Total force 20 B.O.s, 9 Soudanese officers, 6 B.N.C.O.s, 1750 O.R.s of which only 550 were fighting men. 2 B.E.A. & 1 Soudanese Political officers. Weapons included a 2½ lb. Hotchkiss, a Stokes gun. There were some preliminary skirmishes before the force marched on 20.4.18.

Yardley was in command of 6 Coy the Equatorial Bn. (the Equats).

Major Rayne's force consisted of 300 Askaris of the 5 K.A.R., 250 Soudanese Troops and a company from the Equatorial Battalion and they moved north from the Karua River along the banks of Lake Rudolf towards the Lobur Mountain.

The Turkana, who had been joined by many armed Abyssinians, were brought to action on the 6th May when Major Rayne caught a party of raiders driving off cattle. These were broken up by fire from his Hotchkiss gun and the mounted infantry rounded up the stock. The Sudanese were also in action and by mid-May, 141 enemy had been killed and 4,000 or so donkeys and cattle captured.

A further action was fought with Abyssinians, who although brave and daring, were fortunately bad shots.

The troops returned to base on the 18th June.

The Turkana suffered heavy casualties at the Modo Pass where Lolel the chief witch doctor was killed.

Bimbashi White fought an action at Intonga, there was a skirmish at Kaledeke, Nagya.

Capt. Yardley with 70 'Equats,' 30 carriers, 30 levies, 3 Soudanese officers and 6 servants fought a pitched battle at Kangala. Against some 3—400 regular Abyssinians and levies backed by large numbers of Turkana who remained uncommitted.

The Abyssinians attacked with reckless bravery but were beaten off, many of their leaders were killed and next day they fled back to Ethiopia over the Kibish River.

All 3 Soudanese officers were killed and Yardley lost 12 or so killed whilst at least 33 dead Abyssinians were counted.

Many camels, donkeys and cattle were captured.

After the end of the patrol the Northern Turkana were very much left to their own devices and raids continued during the 1920s. Indeed even in the '60/70s the Kenya Army still had to deal with the 'Shifta' on their Borders.

Stigand, the Governor of the Upper Nile Province and White was killed in November, 1919 fighting the Dinka tribe.

The Medal Roll consists of 195 loose pages, and some 67 medals were unclaimed.

Kaimakam R. F. White & H.Q.	21
2 Equatorial Bn. E.A.	513
3 K.A.R.	157
4 K.A.R.	109
5 K.A.R.	93
6 K.A.R.	3
Arab Maxim Battery (2 maxims)	37
9 Sudanese	17
No. 1 Coy M.I. (Egyptian Ranks)	38
Uganda Police	114
Mongolla Province Police	10
Medical Corps	18
Total	1130

CHAPTER TWENTY

Expedition Against the Wakikuyu
& Affiliated Tribes

EAST AFRICA 1902

Army Order No. 102 1908 authorised the African General Service Medal with the clasp 'East Africa 1902, to all officers and men under the command of Lieutenant F. W. O. Maycock, D.S.O., The Suffolk Regiment, composing the Maruka patrol, which reached the Maruka district on the 4th September, 1902, and operated in the district until the 25th October, 1902, both dates inclusive.

This expedition, generally known as the Maruka patrol, was undertaken by the 3 K.A.R. under the command of Lt. F. W. O. Maycock (Suffolk Regiment) against a section of the Kikuyu tribe.

The Kikuyu, now the most powerful tribe in Kenya, were then, as now, concentrated in the highland forest areas around Mount Kenya and the Aberdares, between altitudes of 4,500'–7,500'. The tribe were not then particularly numerous, having recently suffered from smallpox epidemics and the depredations of the Masai. The Kikuyu were essentially forest dwellers, as the pastoral Masai dominated the grasslands around their territories.

The tribe had no firearms and were armed with spears, either bayleaf or Masai pattern, clubs known as Rungus, a long bladed sword called a simi and with bows and arrows.

The Maruka section of the Kikuyu live near Fort Hall in the Kihimbuini district, as they do today, in a country then described by Lt. R. Meinertzhagen (1 Royal Fusiliers) as 'hilly country with dense bush cleared in patches for cultivation.'

Five Indian traders had been murdered by the Kikuyu near the Thika River and the WaKikuyu afterwards refused to submit to Administrative control.

Lt. Maycock therefore led a punitive expedition against them consisting of 5 British officers, 115 rifles of 3 K.A.R., 60 police and 300 Masai levies. The punishment took the form of burning villages and capturing stock, of which about 300 cattle and 2,000 sheep and goats were taken.

Meinertzhagen was already in the area and was informed by a runner of the expedition. He was embarrassed as his caravan was not fitted out for field operations and he had insufficient reserve ammunition. He however captured stock, but met with resistance doing so as his diary records that two of his men were killed and three wounded, whereas the official history states that the total casualties of the patrol were two killed and thirteen wounded.

Maycock informed Meinertzhagen that a white settler had been captured and tortured to death and gave him a free hand to deal with the village. The village was surrounded and all the men and women there, who were dancing around the mutilated body of their victim, which they had used as a latrine, were bayonetted or shot.

MacClean, the Political Officer, apparently refused to give his consent or to interfere with this punishment, an early example of differences of opinion between the Kenya administration and the military.

W.O. 100/394 shows that the following were entitled to the Clasp:—

Lt. F. W. O. Maycock	also '05 & '06
Capt. C. L. Barlow	
Political Officer J. Ainsworth, C.M.G.	
S. L. Hinde	'04 & '06
C. S. Hemsted	
E. B. Horne	
J. W. T. McLellan	'05
Dr. W. J. Radford	
Political Officer A. J. Maclean	
3 K.A.R.	40
E.A. Police	28
Total	78

Strangely Capt. Meinertzhagen does not appear on Roll W.O. 100/394 as entitled to the 1902 clasp, only the 1904. His diary however makes it clear he was in touch with Lt. Maycock who was with him, while Maclean of the Political Dept who was with him is on the roll. Clearly the roll is wrong.

The Medal Rolls show that the police came from Embu, Fort Hall, Kericho, Kiambu, Kisii, Kitui and Machakos police areas, but presumably by the time the roll was prepared the traceable recipients had scattered around the Protectorate.

These figures are considerably less than is indicated from other sources.

EAST AFRICA 1904

Army Order No. 102 of 1908 authorised the clasp 'East Africa 1904' to all officers and men under the command of Captain F. A.

Dickinson, The Duke of Cornwall's Light Infantry, composing the Iraini patrol, which entered the Iraini country on the 13th February 1904 and patrolled it between that date and the 17th March 1904, both dates inclusive.

This was a punitive expedition against another section of the Kikuyu tribe called the Irryeni or Iraini and the Embu tribe as a result of frequent murders of friendly natives.

The expedition was commanded by Captain F. A. Dickinson and comprised 5 officers, 135 rifles of the 3 K.A.R., 60 police and 300 Masai levies. The force operated from bases at Nyeri and Fort Hall into the Gutu and Embu areas. Lt. R. Meinertzhagen's 'Kenya Diary' is critical of the commander and quotes his instructions as 'Never mind my orders, just carry on and don't worry too much, I'll back you up in anything you do.' Lt. Branker left Nyeri with Tate as Political Officer, and Meinertzhagen set out with a Political Officer called Humphrey from Fort Hall with 60 rifles and 250 Masai levies.

In all, over 11,000 stock were captured at the cost of 3 men killed and 33 wounded. The number of enemy killed was estimated at 797 Kikuyu and about 250 Embu. These presumably included an Embu spy who was found disguised as a porter in one of the camps, immediately he was tried by court martial, found guilty and shot only fifteen minutes after capture.

Justice was swift, but women and children were spared. When the Masai levies got out of hand Meinertzhagen records how he himself shot three levies for murdering women and children in a village.

The general opinion was that the Embu had not been taught a severe lesson.

File W.O. 100/394 shows only the following entitlement:—

Capt. Brancker	
Capt. F. A. Dickinson	
Capt. R. Meinertzhagen	
S. L. Hindle	'02 & '06
M. R. Tate	
R. W. Humphrey	
H. B. Bartington	
M. R. McClure	
J. B. Ainsworth	
Political Officer D. W. J. Haywood	
3 K.A.R.	56
E.A. Police	67
Total	133

EAST AFRICA 1906

Army Order No. 102/1908 authorised the clasp 'East Africa 1906,' to all officers and men under the command of Lieutenant F. W. O. Maycock, D.S.O., The Suffolk Regiment, composing the Embo patrol, which entered the Embo country on the 18th June 1906 and operated there until the 19th July 1906, both dates inclusive.
Embu is misspelt in the order.

This was the second punitive expedition commanded by Capt. F. W. O. Maycock, D.S.O.

The Embu tribe, who lived on the lower slopes of Mount Kenya, were reported to be able to deploy between 10,000 and 15,000 spearmen. Their country is much broken up by valleys and heavily afforested.

There are few details available about this patrol except that there were three K.A.R. columns and one of police, which suffered two men killed and fourteen wounded.

Two years later, the Colonial Secretary, Winston Churchill, visited Embu and remarked, 'Yet so peaceful are the tribes now that intertribal fighting has been stopped, that white officers ride freely about their villages never even carrying a pistol. Two young white officers, a civilian and a soldier, presided from this centre of authority over the peace and order of an area as large as an English county.'

Fifty years later however, at the time of the Mau Mau, it would not have been advisable to move around unarmed, in fact, the revolver was standard equipment and carried by all.

File W.O. 100/394 shows the following entitlement:—

Lt. F. W. O. Maycock	'02 & '05
Capt. W. E. M. Barrett	'05
Capt. S. S. Butler	
Capt. W. A. H. Grimshaw	
P.O. S. L. Hinde	'02 & '04
E. B. Horne	'02
C. W. Neligan	
Dr. E. C. Lindsey	
Dr. A. C. Parsons	
C. A. Booth	
Dr. Alexander	
All ranks 3 K.A.R.	184
E.A. Police	90
Total	285

KENYA

Mau Mau Rebellion 1952–56

Unlike 1916, when George V approved a 'new medal' being struck in Army Order 15 of 1955 *the Queen graciously awarded the African General Service Medal (instituted in 1902) for service in operations against the Mau Mau in Kenya in the Central or Southern Provinces, in the Naivasha, Nakuru, or Laikipia Districts of the Rift Valley Province, or in the Isiolo Township or in the Nairobi extra provincial district for a period of 91 days or over since 21st October 1952 to 17th November 1956 inclusive.*

A very wide range of Personnel were eligible for the medal including officers of the 'Prison Staff' and the 'Church of Scotland Huts Committee.' This clasp is therefore the most widely distributed in the series and the Mau Mau rebellion was also the longest single campaign and certainly the most expensive operation conducted in Africa in aid of civil power, as the costs were:—

	£
Borne by the Kenya Government	26,085,428
Grants from Her Majesty's Government	24, 250,000
Interest free loans from Her Majesty's Government	5,250,000
	£ 55,585,428

The Mau Mau activities had more direct effect on more people of all races and classes than any former campaign as nearly everyone over a large portion of Kenya was involved in some way or another by the regulations and restrictions, let alone the active operations.

Casualties were high and the Corfield Report shows:—

	Killed	Wounded
Terrorist Casualties	11,503	1,035
Security Forces		
Europeans	63	101
Asians	3	12
Africans	101	1,469
Loyal Civilians		
Europeans	32	26
Asians	26	36
Africans	1,819	916
Total	13,547	3,595

The fighting solely involved the Kikuyu tribe as the other tribes did not accept their Mau Mau doctrines. The arms used were a few stolen high velocity rifles and pistols, (it is worth noting that of the 1,791 firearms reported as lost, stolen or captured between 1948–1958, 1,635 were recovered), but the Kikuyu relied mainly on their traditional weapon, the 'simi,' or double edged sword, and the 'panga' the universal tool of Kenya. As the Emergency continued, home-made guns were manufactured and used by the Mau Mau. These were highly dangerous for the firer, but did work sometimes, as one of my friends was wounded by a home-made pistol as he entered a native hut when out on patrol, not sufficiently badly however to prevent him disposing of his attacker.

The Security Forces used all the modern weapons available and even four engined heavy bombers were flown to bomb terrorist hideouts in the Aberdare Mountains, much to the indignation of the rhino and elephant, whose home it was, and these animals became steadily more ill tempered and dangerous as time went on.

What caused it all?

It is an oversimplification to say that the Mau Mau movement was a straight forward Freedom Fight as in those days, before 'the wind of change in Africa,' the idea of independence for Kenya seemed absurdly unbelievable, even though this was perhaps the Kikuyu's intention.

The Europeans who had settled in Kenya believed that the country was their home, they thought and acted as if they and their children were to live in Africa for ever. Great pride was taken by them in their achievements in converting virgin land into neat and orderly farms and plantations. They built their own houses and businesses, identified themselves with the country of their adoption and talked of independence from the Colonial Office in Whitehall.

When the troubles started therefore most settlers reacted in defence of their hearth and home.

A Kenya Government pamphlet of 1954 argues that the Kikuyu, due both to external factors, frequent Masai raids, and internal factions, had less sense of personal security than other tribes in Kenya and this perhaps led to the formation of sects and secret societies with oathing ceremonies, to which the Kikuyu are prone.

The Mau Mau creed was anti-European and a typical early type of oath was:—

'If I am sent to bring in the head of an enemy or a European and I fail to do so—May this oath kill me'

'If I fail to steal anything from a European—May this oath kill me'

'If I know of any enemy of our organisation and fail to report it—May this oath kill me'

'If I am ever sent by my leader to do something big for the House of
Kikuyu and I refuse—May this oath kill me'
'If I refuse to help in driving the Europeans from this country—
May this oath kill me'
'If I worship any leader but Jomo Kenyatta—May this oath kill me'
Oathing was also big business as the Oath Administrator was paid a
fee for each initiate and eventually there were seven or so different
grades of oath, for each of which a separate fee was due.

In 1951 and before there had been a certain degree of unrest in the
Kikuyu reserves and this spread to the 'White Highlands,' as the
European farming areas were then designated. There were cases of
arson, attacks on isolated farmers and some murders. The European
settlers began to hear vaguely of Mau Mau, though most of us,
particularly in Nairobi, dismissed the matter as of little import.

This is not the place to record all that went on, much of which
nowadays is better forgotten anyway, but I will tell of some experiences.

In September 1952 Sir Evelyn Baring (later Lord Howick) became
Governor. I attended his swearing-in ceremony outside the Law Courts
in Nairobi. A man standing next to me remarked, 'Poor B..., he doesn't
know what he's let himself in for.' This as it turned out was prophetic,
and years later I mentioned the remark to Lord Howick who laughed
and replied, 'How right he was. I'd no idea how bad things were.'

On 21st October, following a spate of murders of loyal Africans, a
State of Emergency was declared. Jomo Kenyatta and other leaders
were arrested and imprisoned and had no part whatever in subsequent
events in the Emergency.

Many of us had joined the Kenya Police Reserve and undertook
boring and apparently useless duties like manning the local suburban
Police Stations from midnight to midday, waiting for something to
happen, which never did, whilst the regular policeman enjoyed a
peaceful night's sleep in his house. It was distinctly tiresome to
undertake office work in the afternoon and after a few hours sleep go
back to the Police Station for the night.As in the meantime the Kenya
Government did not seem to be doing much about events, some of us
relieved the boredom of the night watches by planning the formation
of a 'Private Army' to go into the European Areas to do something
about chasing the Mau Mau. The particularly unpleasant murder of the
Ruck family, who were butchered in cold blood, gave the final incentive
to march.

Our project was approved by the Kenya Government and a special
unit of some 78 Kenya Police Reserve officers commanded by the late
Lt. Col. D. T. Dobie, D.S.O., of Arnhem fame, was formed and called
'D' Force.

It was essentially a 'city slickers' unit as most of its members came from the Nairobi business community. We had a leading solicitor, a hotelier, land agents and the like.

We provided our own aircraft for reconnaissance and the Government arms and wireless sets. Most members carried their own pistols, some their game rifles. For those who had no firearms of their own, sten guns were issued. Generally the careless handling of firearms caused more near misses than anything the opposition did.

'D' Force then left for the Aberdare Mountains and deployed along the forest boundary above the European farms. At this time there were very few troops in Kenya and our arrival was greatly appreciated by the farmers, but not the Mau Mau, who moved out of the area concerned.

The rear echelon in Nairobi was looked after by my wife who also enrolled as a special policewoman. She and other abandoned wives had to be guarded in our house in Karen by the regular police as the Mau Mau were said to have put a price of £100 on the heads of any member of 'D' Force.

We caught several Kikuyu in what the police would call 'suspicious circumstances' on the forest edge. They were very obdurate and never revealed any information when questioned.

Science was therefore brought to our aid and a closed van with a hidden tape recorder was procured—at that time a modern invention. Two prisoners were placed in the van, and afterwards the tape was played by the intelligence section, which was led by the doyen of Nairobi estate agents. They listened breathlessly—dead silence and one loud fart—the prisoners were strangers to each other and so suspicious they did not speak. Later two friends proved not so reticent, quoth one, 'They are not so clever as they think, they have not found the money sewn into my trousers.' This individual turned out to be a Mau Mau oath administrator and the money helped the mess funds.

At this stage encounters with animals were more frequent than with gangsters. The sound of something approaching the ambush always turned out to be a bush pig or buffalo which would gallop off as soon as they got wind of the patrol. One night march through the forest to search a village at dawn was on a particularly dark night. We were led by trackers, two Africans with spears not unlike the levies used fifty years before. Suddenly the track seemed to run into a blank wall. A torch revealed a large bull elephant standing in the path. He was obviously the most scared as with a trumpet, he about turned and vanished. The trackers bolted back down the column and by the time they reached the far end, which wasn't very long, we were 'under attack by the Mau Mau.' Fortunately the order had been 'nothing up the spout,' so no harm was done.

After a month 'D' Force's duties on the Aberdares were taken over by the K.A.R. and we moved to the Mount Kenya area on the Nanyuki side. Here contact was made with gangs and fire exchanged. Later 'D' Force was moved to Bahati, an African location, then one of the main trouble spots of Nairobi.

By April 1953 more and more regular troops had arrived in Kenya and the older members of the Force returned to the office desks.

What had we achieved? Not much perhaps, except we felt we were doing something, and generally, wherever 'D' Force went, the Mau Mau moved out to safer spots.

After the unit was disbanded it was found that we had a surplus in the Mess Fund, which we decided to spend on a party at the local night club.

It was a good party and as most settlers had strong political views, we decided to call on the Attorney General and then and there express our concern at his handling of affairs. By coincidence some up-country farmers had been a bit more forceful and had telephoned that day to say they were going to burn down the Attorney General's house, and there was therefore an armed guard on his bungalow. The Sergeant in charge was armed with a sten gun and emerged from the darkness in a highly excited state. He executed a type of war dance to the chant of 'My orders are to shoot to kill.' Every time he stamped his foot, the bolt of the sten gun gave a rattle.

However, just as it seemed likely that he would 'bag' half a dozen businessmen, a police 999 car arrived. The Askari, thinking they were our reinforcements, had the police out of the car with their hands up, and it seemed an opportune moment to slip away.

Later we were summoned to the General. a whip-round provided bowler hats, umbrellas and a Rolls Royce for each defaulter. It was silly, but it put the General off his stroke when the motorcade arrived.

Years later, David Dobie became a great friend of the President Jomo Kenyatta, and one day plucked up the courage to ask him 'You know what I did in the Emergency?'

'Of course,' came the somewhat disconcerting reply.

David then said, 'You know that if you were ever in trouble we'd fight for you.'

'Yes,' replied Mzee.

Perhaps this dialogue is a fitting way to end a book on punitive expeditions against freedom fighters.

The medal rolls are not available for inspection, but judging by the very large numbers of Army, auxiliaries, R.A.F. and civilians involved in this campaign, a very considerable number of medals and clasps must have been issued.

APPENDIX NO. 1

H.M. Ships and Weapons

A list of H.M. Ships and other vessels whose crews or boat parties qualified for African Service Medals.

The details shown, which are taken from Jane's Fighting Ships and the card indexes at the National Maritime Museum Greenwich, are incomplete.

ASHANTEE CAMPAIGN 1873–74

PADDLE SLOOPS

H.M.S. Argus (1849), 1630 tons.
H.M.S. Barracuta (1851), 1676 tons, length 100′.

BEACON CLASS GUN VESSEL (COMPOSITE)

H.M.S. Beacon (1867)
570 tons, 155′x25′x7′6″, 9–10½ knots, complement 80.
1 7-in 6½ ton M.L.R., 1 64–pdr. M.L.R., 2 20–pdrs. B.L.

PLOVER CLASS GUN VESSELS (WOODEN)

H.M.S. Seagull (1868)
H.M.S. Bittern (1869)
735 tons, 170′x29′x9′6″, 10 knots, complement 90
1 7-in 6½ ton M.L.R., 2 40–pdrs. B.L. (Armstrong)

ARIEL CLASS COMPOSITE GUNBOATS

H.M.S. Coquette (1871)
H.M.S. Decoy (1871)
H.M.S. Merlin (1871)
430 tons, 125′x22′6″x8′6″, 9½–10½ knots, complement 60
2 64–pdrs. M.L.R., 2 20–pdrs. B.L.

CORVETTES

H.M.S. Active (1869), wood & iron
3980 tons, length 270′ single screw
H.M.S. Encounter (1873), wood
H.M.S. Amethyst (1873), wood 14 64–pdrs. M. L. R., 1970 tons

H.M.S. Druid (1869), wood 1730 tons
H.M.S. Rattlesnake (1866), wood 2431 tons

TROOPSHIPS

H.M.S. Dromedary iron, 1122 tons, single screw, troop and storeship
H.M.S. Himalaya (1854), iron 3 masted. 6 32–pdr. 25 cwt guns 4490 tons, 370'x46'x34' single screw
H.M.S. Simoon (1849), iron frigate converted troopship, 3302 tons, 243'x41'x26'. 18 guns
H.M.S. Tamar (1863), iron, barque rigged, single screw, 2115 tons
H.M.S. Victor Emmanuel (1855), 2 decker wooden single screw 5157 tons, length 230'
S. S. Samaritan

EAST AND WEST AFRICA MEDAL
AFRICAN GENERAL SERVICE MEDAL

Sloops. In 1875 a sloop was a cruising vessel with a complement of between 100 and 200 all ranks

MARINER CLASS SLOOPS (1884)

H.M.S. Acorn   1887–8
H.M.S. Icarus   1887–8
H.M.S. Racer   1891–2, 1892
970 tons, 167'x32'x14', 11 knots, complement 120
8 5–in B.L. 8 Machine–guns

OSPREY CLASS SLOOP

H.M.S. Kingfisher (1879)   Witu 1890
1130 tons, 170'x36'x15', 12 knots, complement 140
2 7–in M.L.R., 4 64–pdrs.

NYMPHE CLASS COMPOSITE SLOOP

H.M.S. Swallow (1885)   1891–2, Witu August 1893
1140 tons, 195'x28'x11'6", 11–13 knots, complement 135
8 5–in B.L., 8 machine guns (later 8 4.7–in Q.F.)

CADMUS CLASS SLOOP (STEEL)

H.M.S. Clio (1903)   Somaliland 1920
H.M.S. Merlin (1901)   Somaliland 1902–04
H.M.S. Odin (1901)   Somaliland 1920
1070 tons, 185'x33'x11'3", 13 knots, complement 120
6 4–in Q.F., 4 3–pdrs.

CRUISERS

COMPOSITE CRUISER, 3 MASTED

H.M.S. Satellite (1881)   Gambia 1894
1420 tons, Length 167', 9–11 knots
2 6–in guns, 6 5–in guns

BRISK CLASS CRUISERS (1885), 3 MASTED

H.M.S. Brisk   Witu 1890
H.M.S. Cossack   Witu 1890, Somaliland 1902–04
H.M.S. Mohawk   Somaliland 1902–04
H.M.S. Porpoise   Somaliland 1902–04
H.M.S. Racoon   Mwele
1770 tons length 225' 13 knots complement 172
8 6–in guns converted Q.F.

SMALL CRUISER STEEL, 3 MASTS

H.M.S. Barham (1890)   Somaliland 1908–10
1830 tons 19 knot twin screw, complement 169
6 4.7–in Q.F., 4 3–pdr. (47 mm)

PELORUS CLASS CRUISERS (1890)

H.M.S. Phœbe   Mwele, Benin 1897
H.M.S. Philomel (1893)   Benin River 1894, Benin 1897,
                        Somaliland 1908–10
H.M.S. Pomone   Somaliland 1902–04
2,575 tons, 320'x49'6"x21', 19 knots complement 190
8 4.7–in, 8 3–pdrs., 4 nordenfelts

CRUISER (1889), SHEATHED & COPPERED

H.M.S. Blanche   Juba River 1893, Witu August 1893
H.M.S. Barossa   Brass River 1895, Mwele, Benin 1897
H.M.S. Blonde   Mwele, Sierra Leone 1898–99
1,580 tons, 233'x35'x16', 15 knots complement 159
6 4.7 in Q.F., 4 3–pdrs., 2 nordenfelts

MEDEA CLASS CRUISER

H.M.S. Magicienne (1888)   Jubaland, Gambia
2,950 tons, length 265' 19 knots complement 218
6 6–in converted, 9 6–pdrs., 3 nordenfelts, 1 9–pdr. boat gun

SMALL CRUISER

H.M.S. Perseus	Somaliland 1902–04
H.M.S. Prosperine	Somaliland 1908–10

2,135 tons, 300'x30'6"x17', 20½ knots complement 224
8 4–in Q.F., 8 3–pdrs.

APOLLO CLASS CRUISER (1891)

H.M.S. Naiad	Somaliland 1902–04
H.M.S. Terpsichore (1890)	Jubaland, Gambia

2,400 tons, 300'x43'x17', 18 knots complement 273
2 6 in., 6 4.7 in, 8 6–pdrs., 1 3–pdr., 4 machine guns

ASTRAEA CLASS CRUISER (1893)

H.M.S. Fox	Sierra Leone 1898–99, Somaliland 1902–04, Somaliland 1908–10
H.M.S. Forte	Gambia 1894, Benin 1897, Gambia

4,300 tons, 320'x49'6"x21', 19½ knots complement 318.
2 6–in, 8 4.7 in Q.F., 8 6–pdrs., 1 3–pdr.

MINERVA CLASS CRUISER

H.M.S. Diana (1895)	Somaliland 1908–10

5600 tons, 364'x54'x23', 19½ knots, complement 450.
11 6–in Q.F., 8 12–pdrs., 7 3–pdrs., 1 12–pdr., 2 maxims.

HERMES CLASS CRUISER (1898), SHEATHED & COPPERED

H.M.S. Highflyer	Somaliland 1902–04
H.M.S. Hyacinthe	Somaliland 1902–03

5,600 tons, 350'x54'x22', 20 knots, complement 450
11 6–in Q.F., 8 3–in Q.F., 6 3–pdr.

EDGAR CLASS CRUISER (1892)

H.M.S. St George	Brass River 1895, Mwele, Benin 1897
H.M.S. Theseus	Benin 1897

7,700 tons, 360'x69'x26', 19 knots, complement 540
2 9.2–in Q.F., 10 6–in Q.F., 12 6–pdrs., 6 3–pdrs., 6 maxims.

FRIGATES, CORVETTES ETC.

H.M.S. Scout (1885)	Jubaland

1,580 tons, 200'x34', 13 knots, complement 147
12 5–in B. L. 40 cwt.

IRON FRIGATE (1873), 3 MAST SQUARE RIG
H.M.S. Raleigh   Gambia 1894
5,200 tons, 298'x49', 13 knots, complement 560
24 guns, 8D+8D+8E

IRON CORVETTE (1875)
H.M.S. Boadicea   Witu 1890
4,140 tons, length 280' 16 guns

3RD CLASS COMPOSITE CORVETTE (1876)
H.M.S. Turquoise   Witu 1890, Benin 1897
2120 tons length 220', 68–pdr. ML

STEEL & IRON CORVETTE (1898)
H.M.S. Conquest   Witu 1890, Benin River 1894, Brass
                           River 1895
2,380 tons length 225'

IRON TROOP & STORE SHIP (1878)
H.M.S. Humber   Witu 1890
1,640 tons length 230'

CONVERTED AIRCRAFT CARRIER
H.M.S. Ark Royal (purchased 1914)
                           Somaliland 1920
7,450 tons, 652'x50'x18', complement RN & RAF 180
4 12–pdrs., no flight deck but cranes for lifting aircraft

GUNBOATS
RENNIE COMPOSITE PADDLE GUNBOAT
H.M.S. Alecto (1882)   1891–2, 1892, Gambia 1894, Benin
                           River 1894, Benin 1897, Sierra Leone
                           1898–99
620 tons, 160'x26'3"x8'10", 8 knots 2 or 4 guns

PIGMY CLASS COMPOSITE GUNBOAT (COST £37,000)
H.M.S. Pigeon (1888)   Witu 1890
755 tons, 165'x30'x11'3", 13 knots, complement 76
6 4–in Q.F.

BRAMBLE CLASS GUNBOAT (STEEL)

H.M.S. Dwarf (1898)   Gambia
710 tons, 180'x33'x8', 13 knots, complement 85
2 4–in Q.F., 4 12–pdrs.

REDBREAST CLASS COMPOSITE GUNBOAT (1898)

H.M.S. Redbreast
H.M.S. Sparrow   1891–2, 1892, Witu August 1893
H.M.S. Thrush   1891–2, 1892, Brass River 1895,
                            Mwele, Gambia, Aro 1901–1902
H.M.S. Widgeon   1891–2, 1892, Gambia 1894, Benin
                            River 1894, Brass River 1895, Mwele,
                            Benin 1897
H.M.S. Magpie   Gambia 1894, Benin 1897
803 tons, 165'x31'x11', 13 knots, complement 76
6 4–in Q.F., 2 3–pdrs.

FROLIC CLASS GUNVESSEL WOODEN

H.M.S. Rifleman (1872)   1887–8
610 tons, 155'x25'x7'9", 11 knots complement 80
1 7–in M.L.R. 6½ ton gun, 1 64–pdrs. M.L.R., 2 20–pdrs. B.L.

TORPEDO GUNBOAT (1894)

H.M.S. Dryad   Somaliland 1902–04
H.M.S. Harrier   Somaliland 1902–04
H.M.S. Hussar   Somaliland 1902–04
1070 tons, 250'x30'6"x13', 18 knots, complement 120
2 4.7–in Q.F., 4 3–pdrs.

LAKE GUNBOATS

H.M.S. Pioneer   Lake Nyassa 1893
H.M.S. Adventure   Lake Nyassa 1893
35 tons, 75'x12' complement 29

STERN WHEEL RIVER GUNBOATS

H.M.S. Herald   Liwondi 1893
H.M.S. Mosquito   Liwondi 1893
75'x12'
4 3–pdrs., later 6. 8 Machine Guns

Side Paddle Cargo Vessel

S.P.V. Dove 25 tons, 65′x14′x1′	Central Africa Medal
S.S. Domira 67 tons, 89′x13½′	Central Africa Medal
S.S. Ilala 21 tons, 55′x10′	Central Africa Medal
H.M.S. Heron	1898

Gunboat

S.S. Jackdaw	Aro 1901–02
S.L. Thrush	1891–2, 1892, Brass River 1894
A Steam Launch, Lt. Comdr. W. M. D'Oyle	

Colonial Steamship

S.S. Countess of Derby	1892, Sierra Leone 1898–99
Colonial Steamer Lily	1891–2

APPENDIX NO. 2

Notes on Some of the Arms Used in Africa

In 1785 2 Trade Guns = 1 Slave = $6

MUSKET

The Brown Bess. ML, black powder .75 calibre and variations thereof known as Dane guns, Tower Muskets, etc.

RIFLES

1851	Delvignie Minie ML, black powder .762 calibre
1863	Enfield ML, black powder .557 calibre Percussion system. Velocity 1200 ft sec (Used greased patch)
1867	Snider Enfield B.L., SA. A conversion of the Enfield into a breach action weapon (breech designed by Jacob Snider of New York and the cartridge and bullet designed by Col Boxer of Woolwich).
1871	Martini=Henry Rifle B.L., SA black powder .45 calibre. A lever action below the small of the butt lowered the breech block. Invented by F. von Martini whilst the barrel was made by A Henry of Edinburgh. Velocity 1350 ft sec. Sighted to 1450 yards.
1887	Lee Metford Rifle B.L. magazine (Mk, 1–8, Mk. 2–10 rds.) Black powder, lead bullet of .303 calibre. Designer of the Bolt, James P. Lee of New York.
1895	Lee Enfield Rifle B.L. magazine as above but cordite and cupronickel coated bullet with a different rifling.
1901	SMLE Mk. 1 (5″ shorter) Sighted 2800 yds. Charger loaded, stocked to muzzle.
1907	SMLE Mk. III Improved charger, blade foresight.
1939	No. 4 rifle, not stocked to muzzle, with different bayonet.

MACHINE GUNS AND AUTOMATICS

Laird Clowes states the following machine guns were used in the Royal Navy in the 19th Century:—

Nordenfelt	.45–in	5 barrelled	Rate of fire 660 rds. per minute
Gardner	.45–in	single barrel	220
		2 barrelled	400
		5 barrelled	650
Gatling	.45–in	10 barrelled	400
Maxim	.45–in calibre		600
	.303–in calibre 600		

The earliest known Nordenfelt machine gun was designed in 1872. It was manually operated and had from 1 to 12 barrels. The rounds were fed from a hopper. The 3 barrel version was adopted by British Army in 1884. Cyclic rate 350 rds. per minute. In 1890 Nordenfelt combined with Maxims as Maxim–Nordenfelt. Nordenfelts also made up to 1–pdr. automatics.

The Gardner was an American invention with a single barrel in final form firing 320 rds. per minute.

The Gatling was designed in 1861 by Dr. Richard Gatling as a 6 barrelled, crank operated gun, adopted by the U.S. Army in 1866 as a .5–in calibre weapon. It was later adapted for gas firing with a firing rate of 3000 rds. per minute. It was obsolete in 1911. It was used by the British in Africa.

The Maxim was invented in 1884, was belt fed and water cooled. As a .45–in calibre weapon it was adopted by the British Army later as a .303–in and the Vickers Mark 1 is only a modification of the original name adopted 1915.

Another machine gun used was the gas operated Hotchkiss (1895) with strip feed.

Apart from the more general use of shells as projectiles Naval Gunnery virtually stagnated up to the Crimean War.

In 1858 the Committee on Rifled Cannon recommended the introduction of Rifled Breechloading Armstrong guns with Hexagonal rifling and screw breeches. By 1860 in the Navy 9–pdr., 12–pdr., 20–pdr., 40–pdr. and heavy 7–in breechloading guns were in use.

These Armstrong guns were unsatisfactory and the 40–pdr. and 7–in quickly abandoned and in the mid–1860s the Navy reverted to muzzleloaders and adopted R.M.L. 7–in, 8–in, 10–in, 11–in, 12.5–in and 16–in. These muzzleloaders were standard equipment until 1881 when more efficient manufacturing processes allowed the reintroduction of breech loading guns, but even in 1900 some ships still retained muzzle loaders.

Q.F. guns with fixed ammunition were first introduced in 1881.

The Woolwich Armstrong Breechloaders included 12 pdrs., 4–in, 6–in, 8–in, 9.2–in, 12–in, 13.5–in and 16¼–in. Propellants changed to cordite in 1895–6.

Bibliography

BRITISH BATTLES AND MEDALS; Maj. L. L. Gordon; Spink & Son Ltd.

THE HISTORY OF THE ROYAL WEST AFRICA FRONTIER FORCE; Haywood and Clark; Gale & Polden

THE KINGS AFRICAN RIFLES; Lt. Col. H. Moyse-Bartlett; Gale & Polden

THE ROYAL NAVY, Vol. 7; Sir W. Laird; Clowes

FRONTIER & OVERSEAS EXPEDITIONS FROM INDIA, Vol. VI; Government Printing Press, Calcutta 1911

ALFRED YARROW, HIS LIFE & WORKS; Eleanor Barnes; Edward Arnold & Co

LAKE MALAWI STEAMERS; P. A. Cole King

THE PSYCHOLOGY OF MAU MAU; Dr. J. C. Carrotlier; 1954 Government Printers, Nairobi

THE ORIGINS AND GROWTH OF MAU MAU; F. D. Corfield; 1960 Government Printers, Nairobi

KENYA DIARY 1902–1926; Col. R. Mainertzhagen; Oliver & Boyd

ARMY DIARY 1899–1926; Col. R. Mainertzhagen; Oliver & Boyd

THE IRON SNAKE; Ronald Hardy

SCRAMBLE FOR AFRICA; Anthony Nutting

SEND A GUNBOAT; Preston and Major

K.A.R.; W. Lloyd Jones; Arrowsmith

FIGHTING THE SLAVE HUNTERS IN CENTRAL AFRICA; Alfred J. Swann; Frank Cass & Co. Ltd.

THE ENCYCLOPEDIA OF BRITISH EMPIRE POSTAGE STAMPS; Robson Lowe

MORE BANDOBAST; Snaffles

LUGARD, Vol. 1 and Vol. 2; Margery Perham

AFRICAN INCIDENTS; Major A. B. Thurston; John Murray London 1900

THE LAND OF NILE SPRINGS; Colonel Sir Henry Colvile, K.C.M.G., C.B.; Edward Arnold London 1895

RODDY OWEN (Brevet Major Lancashire Fusiliers, D.S.O.) by his sister Mai Bovill & B. R. Askwith; John Murray London 1897

MAJOR JOHN YARDLEY D.S.O.; Paragon or Eddies in Equatoria; J. M. Dent & Sons London 1931

TRAVELS & RESEARCHES AMONG THE LAKES & MOUNTAINS OF EASTERN & CENTRAL AFRICA; J. F. Elton; John Murray 1879
THE BENIN MASSACRE; Capt. Alan Boisragon; Methuen 1897
AFTER LIVINGSTONE; Fred M. Moir; Hodder & Stoughton
THE RISE OF OUR AFRICAN EMPIRE; F. D. Lugard
MY AFRICAN JOURNEY; Winston Churchill
INFANTRY UNIFORMS; R. & W. Wilkinson; Latham
ENCYCLOPEDIA OF FIRE ARMS; H. L. Peterson
TRENCHARD; Andrew Boyle
CAMPAIGNING IN THE UPPER NILE AND NIGER; Lt. Seymour Vandeleur
ADVENTURES IN NYASSALAND; L. Montieth Fotheringham, 1891
THE LUNATIC EXPRESS; Charles Miller
NIGERIA, A HISTORY; John Hatch
A HISTORY OF LORD LUGARD'S CONQUEST OF HAUSALAND; D. J. M. Muffett
SMALL WARS, THEIR PRINCIPLES & PRACTICE; C. E. Callwell; H.M.S.O. 1906
STORY OF THE RHODESIAS & NYASSALAND; A. J. Hanno; Faber
THE MODERN HISTORY OF SOMALILAND; I. M. Lewis
KINGDOMS OF YORUBA; R. S. Smith
IMPERIAL SUNSET; Maj. Gen. James Lunt; Macdonell 1981
THE ADVANCE OF AN AFRICAN EMPIRE; Wallis

Index of Clasps

Index

www.ingramcontent.com/pod-product-compliance
Lightning Source LLC
Chambersburg PA
CBHW080512030726
47592CB00012B/3326